# Studies in Fuzziness and Soft Computing

Volume 398

The series “Studies in Fuzziness and Soft Computing” contains publications on various topics in the area of soft computing, which include fuzzy sets, rough sets, neural networks, evolutionary computation, probabilistic and evidential reasoning, multi-valued logic, and related fields. The publications within “Studies in Fuzziness and Soft Computing” are primarily monographs and edited volumes. They cover significant recent developments in the field, both of a foundational and applicable character. An important feature of the series is its short publication time and world-wide distribution. This permits a rapid and broad dissemination of research results.

Indexed by ISI, DBLP and Ulrichs, SCOPUS, Zentralblatt Math, GeoRef, Current Mathematical Publications, IngentaConnect, MetaPress and Springerlink. The books of the series are submitted for indexing to Web of Science.

More information about this series at http://www.springer.com/series/2941

Muhammet Gul · Suleyman Mete ·
Faruk Serin · Erkan Celik

# Fine–Kinney-Based Fuzzy Multi-criteria Occupational Risk Assessment

## Approaches, Case Studies and Python Applications

Muhammet Gul
Department of Emergency Aid
and Disaster Management
Munzur University
Tunceli, Turkey

Faruk Serin
Department of Computer Engineering
Munzur University
Tunceli, Turkey

Suleyman Mete
Department of Industrial Engineering
Gaziantep University
Gaziantep, Turkey

Erkan Celik
Department of Transportation
and Logistics
Istanbul University
Istanbul, Turkey

ISSN 1434-9922 ISSN 1860-0808 (electronic)
Studies in Fuzziness and Soft Computing
ISBN 978-3-030-52150-9 ISBN 978-3-030-52148-6 (eBook)
https://doi.org/10.1007/978-3-030-52148-6



This Springer imprint is published by the registered company Springer Nature Switzerland AG
The registered company address is: Gewerbestrasse 11, 6330 Cham, Switzerland

*To my dad* **Ali**, *mom* ***Fevziye***, *wife* ***Tuğba*** *and children,* ***Ahmet Faruk*** *and* **Zeynep Sena**

*Muhammet Gul*

*To my dad* ***Mahmut***, *mom* ***Latife***, *wife* ***Hayriye*** *and children,* ***Yusuf*** *and Latife Nisa*

*Suleyman Mete*

*To* ***Sabri, Abdulkadir, Zülfiye, Hediye, Emine, Muhammed Rohat, Alya Nil***

*Faruk Serin*

*To my dad* ***Cemil***, *mom* ***Zeliha***, *wife* ***Canan***, *and son* ***Muhammed***

*Erkan Celik*

# Preface

Since its development as an occupational risk assessment method in the 1970s [1, 2], Fine–Kinney method has been extensively applied to a diversity number of safety risk assessment problems of various industries. It is considered as a systematic methodology presenting a formula for calculating the risk due to a hazard. Classically, safety risk assessment via Fine–Kinney is carried out by a score which is the arithmetic product of probability (P), exposure (E) and consequence (C). Calculated risk scores help decision-makers prioritize corrective and preventive actions. This simple, easily understood and useful method is practically used by small and medium enterprises for their risk analysis processes. On the other side, in the academic knowledge, Fine–Kinney method is yet applied to various areas recently although it has several drawbacks. Since this method aims to prioritize the hazards and associated risks depending on three different parameters, the problem that this method tries to solve can be considered as a classic multi-criteria decision-making (MCDM) problem.

MCDM methods are frequently applied to occupational risk assessment problems by many scholars. In MCDM methods with crisp and precise data, the performance ratings and the weights of the decision criteria are known precisely and indicated by crisp numbers. However, many real-world problems involve uncertain data and one cannot assume the knowledge and judgments of the experts to be precise. Hence, fuzzy-based MCDM methods are proposed to reflect types and degrees of uncertainties better than classical ones. At this point, this book offers a number of approaches to Fine–Kinney-based multi-criteria occupational risk assessment.

Occupational risk assessment is a sub-process of risk management for evaluating of the risks arising from a hazard, considering the required control measures and deciding whether the risks are acceptable or not. It includes the determination of a quantitative or qualitative value for the risk. Several approaches to perform occupational risk assessment are available in the literature ranging from expert to participatory methodologies and from simple to complex methods. The Fine–Kinney method is an exhaustive occupational risk assessment method for quantitative evaluation of the hazards.

The book provides valuable insights into useful Fine–Kinney-based fuzzy multi-criteria occupational risk assessment approaches, case studies that can show the applicability of each proposed approach and Python coding of each proposed approach that can be useful for stakeholders to easy implement those to their risk assessment process.

The book is organized to include nine chapters. Chapter 1 contains the basics of Fine–Kinney method with its implementing procedure, its basic terminology and drawbacks. In addition to this necessary information, a state-of-the-art review is provided including its extensions by fuzzy sets. Graphical results obtained from the review are demonstrated to show the current state-of-the-art. Future work suggestions are also included in the chapter to show the possible gaps and possible opportunities.

Chapter 2 introduces a Fine–Kinney-based occupational risk assessment using fuzzy best and worst method (F-BWM) and fuzzy multi-attribute ideal real comparative analysis (F-MAIRCA). In this chapter, instead of crisp numbers, triangular fuzzy numbers that reflect the uncertainty well in real-world problems are used in integration with BWM method in determining Fine–Kinney risk parameters. The risks are then prioritized by F-MAIRCA. A case study for the occupational risk assessment of raw mill processes in a cement factory has been conducted to demonstrate the feasibility of the approach, and besides this case study, a comparative study has also been conducted to test the validity of the proposed approach. Python implementation has been done to help stakeholders easy model its safety problem by adapting this novel approach.

Chapter 3, an improved Fine–Kinney occupational risk assessment approach is proposed using a well-known MCDM method "TOPSIS" under interval type-2 fuzzy set concept. It is defined as technique for order preference by similarity to ideal solution by Hwang and Yoon (1980). It is based on separation from ideal and anti-ideal solution concept. Interval type-2 fuzzy set is an improved version of type-1 fuzzy set. It is also special version of a general type-2 fuzzy set. Since general type-2 fuzzy systems contain complex computational operations, they have not easily applied to real-world problems such as occupational risk assessment. Interval type-2 fuzzy sets are the most frequently used type-2 fuzzy sets due to their ability in handling more uncertainty and producing more accurate and solid results. The Fine–Kinney concept is merged with interval type-2 fuzzy set concept and TOPSIS for the first time through the literature. To demonstrate the applicability of the proposed approach, a case study is carried out in a chrome plating unit of a gun factory. Some beneficial validation and sensitivity analysis are also performed. Finally, as a creative contribution of our book, the implementation of the proposed approach in Python is performed.

Chapter 4 aims at adaptation of Fine–Kinney occupational risk assessment concept into VIKOR multi-attribute decision making method with interval-valued Pythagorean fuzzy set. The classical fuzzy set theory has been improved by proposing a number of extended versions. One of them is Pythagorean fuzzy set. In this chapter, we use this type of fuzzy set with VIKOR since it reflects uncertainty in occupational risk assessment and decision-making better than other fuzzy extensions. To demonstrate the proposed approach applicability, a case study regarding the activities of surface treatment area in a chrome plating unit of a gun factory is performed. Some additional analysis to test the solidity and validity of the approach is executed. Finally, the Python codes in the implementation of the proposed approach are given for scholars and practitioners for usage in further studies.

Chapter 5 applies a novel occupational risk assessment approach which merges the TODIM with Fine–Kinney method under intuitionistic fuzzy set concept. Risk parameters of Fine–Kinney and OHS experts are weighted by an intuitionistic fuzzy weighted averaging (IFWA) aggregation operator. Hence, hazards are quantitatively evaluated and prioritized using the proposed approach. To illustrate the novel risk assessment approach, processes of the gun and rifle assembly line of a factory are handled. A comprehensive risk assessment is carried out to improve operational safety and reliability in the industry. We adapt intuitionistic fuzzy sets in the existed study since they reflect uncertainty with a aid of their membership and non-membership functions in decision-making better than classical fuzzy extensions. An additional sensitivity analysis by changing the attenuation parameter of TODIM is performed to test the validity of the approach. Finally, the Python codes in the implementation of the proposed approach are given provided as well.

Chapter 6, a Fine–Kinney-based occupational risk assessment using hexagonal fuzzy multi-objective optimization by ratio analysis (HFMULTIMOORA) is handled. Hexagonal fuzzy numbers (HFNs) can be used as a proficient logic to simplify understanding of ambiguity information. HFNs present the usual information in a comprehensive way and also the ambiguity section can be exemplified in a reasonable way. In this chapter, we proposed an improved Fine–Kinney occupational risk assessment approach using an integration of MULTIMOORA and HFNs. To show the applicability of the novel approach, a case study of risk assessment of a raw mill in cement plant was provided. Comparative analysis using two aggregation tools as reciprocal rank method and dominance theory is carried out. Finally, the Python implementation of the proposed approach is implemented to be effective for those concerned in the future.

Chapter 7 includes an improved Fine–Kinney approach using neutrosophic sets. Neutrosophic sets reflect uncertainty and vagueness in real-world problems better than classical fuzzy set theory. It takes into three decision-making situations consideration called indeterminacy, truthiness, and falsity. In Zadeh's traditional fuzzy set theory, there is just membership function fuzzy set degree. But, in neutrosophic environment, it considers three membership functions. Unlike intuitionistic fuzzy sets, an indeterminacy degree is considered. In this chapter, we applied a special form of neutrosophic set as single-valued neutrosophic set (SVNs) with TOPSIS under the concept of Fine–Kinney occupational risk assessment. Since the mere TOPSIS has failed to handle imprecise and vague information which usually exist in real-world problems, we follow integration of SVNs and TOPSIS. To demonstrate the applicability of the novel approach, a case study of risk assessment of a wind turbine in times of operation was provided. Comparative analysis with some similar approaches and sensitivity analysis by changing the weights of Fine–Kinney parameters are carried out. Finally, the Python implementation of the proposed approach is executed as in other chapters.

Chapter 8, we improved Fine–Kinney occupational risk assessment approach with interval type-2 fuzzy QUALIFLEX (IT2FQAULIFLEX). QAULIFLEX is an outranking multi-attribute decision-making method proposed by an extension of Paelinck's generalized Jacquet-Lagreze's permutation method. Similar to other outranking solution-based approaches, it considers the solution which is comparison of hazards. In this chapter, we adapted the interval type-2 fuzzy sets (IT2FSs) into QAULIFLEX as it reflects the uncertainty well in decision-making. IT2FQAULIFLEX algorithm under the Fine–Kinney concept provides a useful and solid approach to the occupational health and safety risk assessment. In addition to proposing this new approach, a case study is performed in chrome plating unit. A validation is also performed in this study. Finally, the proposed approach is implemented in Python.

Final chapter (Chap. 9) provides an improved Fine–Kinney occupational risk assessment approach with interval type-2 fuzzy VIKOR (IT2FVIKOR). VIKOR is a compromise multi-attribute decision-making method proposed by Opricovic (1998). Similar to other compromised solution-based approaches, it considers the solution which is closest to the ideal. In this chapter, we adapted the interval type-2 fuzzy sets (IT2FSs) into VIKOR as it reflects the uncertainty well in decision-making. IT2FVIKOR algorithm under the Fine–Kinney concept provides a useful and solid approach to the occupational health and safety risk assessment. In addition to proposing this new approach, a case study is performed in a gun and rifle barrel external surface oxidation and colouring unit of a gun factory. A validation and a sensitivity analysis are also attached to this study. Finally, the proposed approach is implemented in Python.

This book will be one of the most important guidance books of professionals and researchers working in the field of occupational risk management. It can be considered as a guide document for how an industrial organization proactively identifies, manages and mitigates the risk of patient hazard. It also aims to become a valuable reference book for postgraduate and undergraduate students.

Finally, we, as the four editors of this book, are grateful to our families for their constant love, patience and support. Without their unique support, we would not have been able to complete this book. With the best wishes that the book will be useful to all concerned.

Tunceli, Turkey — Muhammet Gul
Gaziantep, Turkey — Suleyman Mete
Tunceli, Turkey — Faruk Serin
Istanbul, Turkey — Erkan Celik
March 2020

## References

1. Fine, W. T. (1971). Mathematical evaluations for controlling hazards. *Journal of Safety Research*, *3*(4),157–166.
2. Kinney, G. F., & Wiruth, A. D. (1976). *Practical risk analysis for safety management* (pp. 1–20). Naval Weapons Center.

# Contents

# Chapter 1
# Fine–Kinney Occupational Risk Assessment Method and Its Extensions by Fuzzy Sets: A State-of-the-Art Review

**Abstract** The Fine–Kinney method (Fine in J Saf Res 3:157–166, 1971; Kinney and Wiruth in Practical risk analysis for safety management. Naval Weapons Center, pp 1–20, 1976), which was first introduced as an occupational health and safety risk analysis tool in the 1970s, is a systematic methodology that provides a mathematical formula for calculating the risk that arises due to a specified hazard. In the traditional version of Fine–Kinney as suggested in its original version, a risk score (RS) is calculated as a result of mathematical multiplication of probability (P), exposure (E), and consequence (C) parameters. These calculated risk scores are used to establish priorities for the corrective efforts in order to eliminate risks or reduce their effects to a reasonable level. This simple and useful method is preferred and implemented by small and medium-sized enterprises. In the academic literature, it has been applied for many risk analysis problems, although it includes several drawbacks recently revealed. In this method, no weight assignment is made for each risk parameter. Also, it is hard to assess consequence, exposure, and probability, precisely. Multi-criteria decision making (MCDM) is a pool of methods used in occupational health and safety risk analysis both by international standard-setting organizations and scholars from the literature. In classical MCDM methods, performance values and weights of decision criteria are known precisely and are specified with crisp numbers. However, many real-world problems contain uncertainties, and the knowledge and judgment of experts cannot be expressed precisely. Fuzzy-based MCDM methods, which are developed to reflect types and degrees of uncertainties better, produce more accurate results compared to classical methods. In this chapter, we first present the basics of Fine–Kinney method, including its implementing procedure, basic terminology, and drawbacks. Then, we provide a state-of-the-art review of Fine–Kinney occupational risk assessment method and its extensions by fuzzy sets. Graphical results obtained from the review are demonstrated to show the current state. Future work suggestions are also included to the chapter to show the possible gaps and possible opportunities.



M. Gul et al., *Fine–Kinney-Based Fuzzy Multi-criteria Occupational Risk Assessment*, Studies in Fuzziness and Soft Computing 398,
https://doi.org/10.1007/978-3-030-52148-6_1

## 1.1 Classical Fine–Kinney Method

This method is developed for risk assessment problems. It is a simple technique and can be easily adopted by safety stakeholders in any industry from manufacturing to service. In this method, the RS is computed by multiplying the parameters of $P$, $E$, and $C$. The formula of risk score is created as follows: $RS = P \times E \times C$. The ratings regarding these three parameters are demonstrated in Fig. 1.1. Clusters of risks are also reachable by this figure. $P$ is the risk probability of the hazard-event. $E$ is described as the frequency of occurrence of the hazard-event. $C$ is defined as the most likely result of an undesirable event [1].

Many variations of Fine–Kinney method are proposed over time. Some terms commonly used in conducting a Fine–Kinney occupational risk assessment are defined as follows:

*Hazard*: It is defined as any unsafe condition or potential source of an undesirable accident event. For example, welding in a tank; defective brakes on a vehicle; transport of heavy loads by hand.

*Probability*: It is defined as the likelihood of an accident or damage when the hazard occurs.

*Exposure*: It is defined as the frequency of occurrence of the hazard-event.

*Consequence*: It is defined as the most probable result of a potential undesirable accident event, including injuries and property damages.

*Risk*: It is defined as the chance of something to have an impact on the target. It is based on probability, exposure, and consequence.

*Risk management*: It is defined as a process including determining the context, identifying the risks, analyzing the risks, evaluating the risks, treating the risks, continuous monitoring and reviewing, and communicating and consultation.

For a more detailed vocabulary list of Fine–Kinney, please see Fine [1], Kinney and Wiruth [2], and Dickson [3].

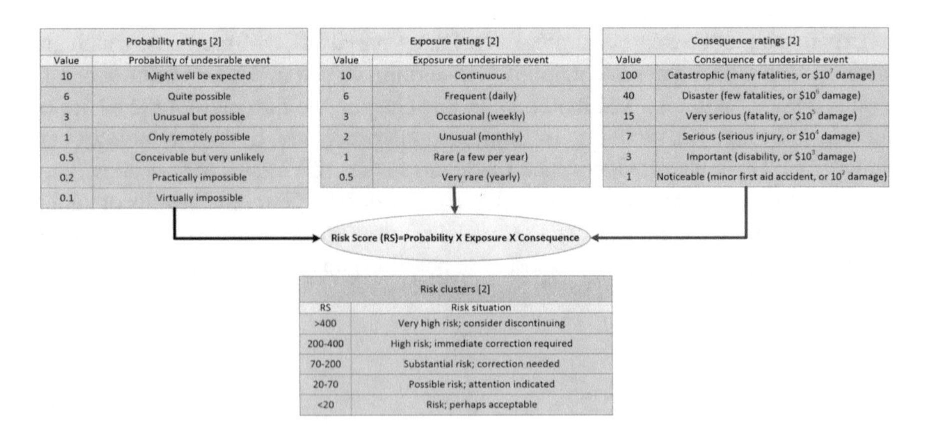

| Probability ratings [2] | |
|---|---|
| Value | Probability of undesirable event |
| 10 | Might well be expected |
| 6 | Quite possible |
| 3 | Unusual but possible |
| 1 | Only remotely possible |
| 0.5 | Conceivable but very unlikely |
| 0.2 | Practically impossible |
| 0.1 | Virtually impossible |

| Exposure ratings [2] | |
|---|---|
| Value | Exposure of undesirable event |
| 10 | Continuous |
| 6 | Frequent (daily) |
| 3 | Occasional (weekly) |
| 2 | Unusual (monthly) |
| 1 | Rare (a few per year) |
| 0.5 | Very rare (yearly) |

| Consequence ratings [2] | |
|---|---|
| Value | Consequence of undesirable event |
| 100 | Catastrophic (many fatalities, or $10$^{7}$ damage) |
| 40 | Disaster (few fatalities, or $10$^{6}$ damage) |
| 15 | Very serious (fatality, or $10$^{5}$ damage) |
| 7 | Serious (serious injury, or $10$^{4}$ damage) |
| 3 | Important (disability, or $10$^{3}$ damage) |
| 1 | Noticeable (minor first aid accident, or $10^{2}$ damage) |



| Risk clusters [2] | |
|---|---|
| RS | Risk situation |
| >400 | Very high risk; consider discontinuing |
| 200-400 | High risk; immediate correction required |
| 70-200 | Substantial risk; correction needed |
| 20-70 | Possible risk; attention indicated |
| <20 | Risk; perhaps acceptable |

**Fig. 1.1** Flowchart of Fine–Kinney method

**Table 1.1** A sample worksheet format for Fine–Kinney occupational risk assessment

| Project Name | | Doc No: | | Revision No | |
|---|---|---|---|---|---|
| Doc Name | | Prepared Date: | | Page | |

| Activity and Location | Hazard and/or occurring of risk | Consequence of hazard | Who Effected | Degree of Risk | | | | Result | Control Measure(s) to be implemented | Residual Degree of Risk | | | | Person Responsible for Control Measure(s) |
|---|---|---|---|---|---|---|---|---|---|---|---|---|---|---|
| | | | | Probability | Exposure | Consequence | Risk Degree (Priority) | | | Probability | Exposure | Consequence | Risk Degree (Priority) | |
| | | | | | | | | | | | | | | |

To perform an occupational risk assessment by Fine–Kinney method, the following steps must be followed:

**Step 1**: List the accident-sequence of events that could result in the undesired consequences.

**Step 2**: Determine values for elements of formula (probability, exposure, and consequence). The values are given in Fig. 1.1.

**Step 3**: Substitute into the formula and determine the RS.

**Step 4**: Prioritize the hazards for preventive actions. The hazards are prioritized in terms of the RSs in decreasing order. Hereafter, suggested control measures for the high-risk hazards should be taken into consideration.

**Step 5**: Prepare a Fine–Kinney occupational risk assessment report by summarizing the analysis results. A sample worksheet about Fine–Kinney is demonstrated in Table 1.1.

**Step 6**: Calculate the revised RSs to reduce or eliminate risks posed by hazards. After the proposed control measures and corrective actions are put in place, the risk assessment team should reassess each of the risk sequences for P, E, and C. This risk reassessment is very important because it shows the picture of how much the risk associated with each hazard has been improved (reduced to an acceptable level or completely eliminated in the long term).

## 1.2 Drawbacks of Fine–Kinney Method

The classical Fine–Kinney method is considered as a beneficial occupational risk assessment tool that aids to assess potential hazards and associated risks. However, in the current academic knowledge, there exist a number of drawbacks for the classical RS calculation of this method. Some of them are summarized as follows [4, 5]:

- The relative importance (weight) values of P, E, and C are ignored in the RS calculation.
- Various combinations of P, E, and C may result in the same value of RS, but their meanings under risk implications may be completely different. The situation may cause not notice of some high-risk hazards.
- The calculated RS is questionable and does not lie on an exact scientific source. There is no scientific and realistic justification for why P, E, and C should be multiplied to obtain RS.
- The interrelationships between hazards and causes of hazard are not discussed sufficiently.
- The parameters of "P, E, and C" are analyzed based on a scale with discrete measurement values. However, the calculation of the product is meaningless on an ordinal scale.
- The RS is computed regarding merely three risk factors, mainly in terms of occupational safety. Some other crucial parameters such as cost, prevention, sensitivity to non-utilization of personal protective equipment, sensitivity to non-execution of maintenance, un-detection are unfortunately out-of-scope.
- Incorrect definition of probability, the necessity to base the accident for one reason, estimation of exposure factor, and number of workers. Refer to [4] for more satisfying arguments.
- The mathematical formula of the RS calculation is extremely sensitive to the change of the values of risk parameters. Small changes in risk parameter values lead to very different changes in the RS value.

## 1.3 Methodology of the State-of-the-Art Review

Regarding drawbacks in the classical Fine–Kinney, several studies have been performed on the improvement of Fine–Kinney using MCDM methods and their fuzzy extensions. In this chapter, we provide a state-of-the-art review considering the deficiencies of the classical Fine–Kinney method based on MCDM methods and fuzzy extensions. A total of 19 articles from the international journals published between 2008 and 2019 is gathered and analyzed. In this section, the results of the state-of-the-art review are demonstrated. For the review, we have searched by suitable keywords of "Fine–Kinney", "risk assessment", "occupational health and safety (OHS)", "proportional risk assessment" in various combinations. The searching work is carried out through Google Scholar. We have excluded the thesis, reports, and

**Table 1.2** Categorization of Fine–Kinney papers

| Category | MCDM methods used | References |
|---|---|---|
| Pairwise comparison MCDM methods | AHP | Kokangül et al. [6] |
| | AHP | Yilmaz and Ozcan [23] |
| | PFAHP | Karasan et al. [7] |
| | PFAHP | Ilbahar et al. [8] |
| | FAHP | Zhang et al. [24] |
| Hybrid MCDM methods | FAHP, FVIKOR | Gul et al. [9] |
| | FAHP, FVIKOR | Gul et al. [10] |
| | FAHP, FVIKOR | Gul et al. [11] |
| Without MCDM methods | – | Oturakçı et al. [13] |
| | | Marhavilas and Koulouriotis [14] |
| | | Gurcanli et al. [15] |
| | | Netro et al. [16] |
| | | Korkmaz et al. [17] |
| | | Makajic-Nikolic et al. [18] |
| | | Birgören [19] |
| | | Gul and Celik [20] |
| | | Supciller and Abali [21] |
| | | Oturakçı and Dağsuyu [22] |
| Other MCDM methods | MULTIMOORA, Choquet integral | Wang et al. [12] |

unpublished working papers from the review. A variety of MCDM-based occupational risk assessment models are existed in the academic knowledge to improve and extend Fine–Kinney. The scanned and reviewed papers are merged under four categories: (1) Pairwise comparison MCDM methods; (2) Hybrid MCDM methods; (3) Without MCDM methods; (4) Other MCDM methods. These papers are investigated in detail in the next section. The categorization of these papers is shown in Table 1.2.

## 1.4 Fine–Kinney Using MCDM Methods and Fuzzy Sets

Regarding the first category models, Kokangül et al. [6] conducted a risk assessment for a manufacturing facility where hazards were identified based on expert experience. Historical accident records were classified and priorities were determined for each class using the AHP method. Hazards and associated risks were also evaluated using the Fine–Kinney method. The risk classes in both the Fine Kinney risk assessment and AHP were revealed and as a result of the study, it was concluded that the risk class measurement in Fine–Kinney can be used in the results obtained from the

AHP method. An extended version of AHP was also used in two studies performed by Karasan et al. [7] and Ilbahar et al. [8]. Karasan et al. [7] proposed a novel approach, Safety and Critical Effect Analysis (SCEA), considering its extension with Pythagorean fuzzy sets for the first time in the risk assessment literature. In the practical case of [7], hazards identified for an excavation activity were assessed using both the SCEA and Pythagorean fuzzy sets. The results obtained were compared with FMEA and Fine–Kinney methods and it was found that the proposed methods provided a more reliable and informative output based on all necessary parameters. In Ilbahar et al. [8], a novel integrated approach, Pythagorean Fuzzy Proportional Risk Assessment (PFPRA), including Fine Kinney, Pythagorean fuzzy AHP, and a fuzzy inference system was used in the OHS field. Wilma and Ozcan [23] performed a risk analysis and ranking application for lifting vehicles used in construction sites integrated AHP and Fine–Kinney. Zhang et al. [24] proposed a combined method including FAHP and Fine–Kinney for airport operation situation risk assessment.

Besides the single approaches, there exist many hybrid MCDM methods in Fine–Kinney-based risk assessment literature. Gul et al. [9] presented a hybrid risk-based approach for the maritime industry by adopting the Fuzzy Analytic Hierarchy Process (FAHP) with fuzzy VIKOR (FVIKOR) methods under Fine–Kinney approach. In another two studies by Gul et al. [10, 11], both FAHP and FVIKOR were applied. Other MCDM methods such as MULTIMOORA and Choquet integral were used together in a Fine–Kinney risk assessment model in [12].

In many times, Fine–Kinney is studied without MCDM methods [13–22]. Oturakçı et al. [13] made an improvement in the pure Fine–Kinney method. That is, alternative scales of Fine–Kinney were re-created for the parameters of “probability” and “frequency”. It was observed that risk scores, which are obtained from the new approach, were more sensitive than classical Fine–Kinney Methods’ risk scores. In [14], comparison between classical $5 \times 5$ matrix method and Fine–Kinney by a case study in the aluminum extrusion industry. Netro et al. [16] adapted the Fine–Kinney method to supply chain risk management. On the other hand, Korkmaz et al. [17] used Fine–Kinney and 5S methodology for risk assessment and management process of a plastic injection and mounting departments of a factory. From a different application perspective, Makajic-Nikolic et al. [18] applied the Fine–Kinney method, which was originally developed for OHS analysis, for the risk assessment of terrorism at tourism destinations. Birgören [19] contributed to Fine–Kinney methodologically. The author studied the calculation challenges and solution suggestions for risk parameters in the Fine Kinney risk analysis method. Gul and Celik [20] merged Fine–Kinney method with a fuzzy-rule-based expert system. The case of risk assessment in the rail transportation system of Istanbul (Turkey) is handled and a comparison with the classical Fine–Kinney method was discussed. In [21], the authors preferred to mention the name of Fine–Kinney method as a proportional risk assessment technique (PRAT). They proposed a risk analysis with the fuzzy PRAT for the first time to overcome the drawbacks of the conventional PRAT method. Oturakçı and Dağsuyu [22] presented fuzzy Fine–Kinney method using a rule-based expert system.

## 1.5 Analysis Results of the Review

Based on the reviewed papers on the improvement of Fine–Kinney using MCDM methods and their fuzzy extensions (19 papers), a short analysis is conducted in this chapter regarding the annual publication trend, the version of fuzzy sets adopted in Fine–Kinney, the journals in which the articles are published, the application areas of Fine–Kinney, MCDM methods used in the paper, availability of a comparative analysis, availability of a sensitivity analysis, and type of application. The analysis results are given in the following figures (Figs. 1.2, 1.3, 1.4, 1.5, and 1.6).

A total of 19 papers were analyzed in this chapter. While most of them 17 (89%) belong to journal articles, 2 (11%) papers are presented at selected conference proceedings. When we look at the distribution of the studies in time, it can be seen

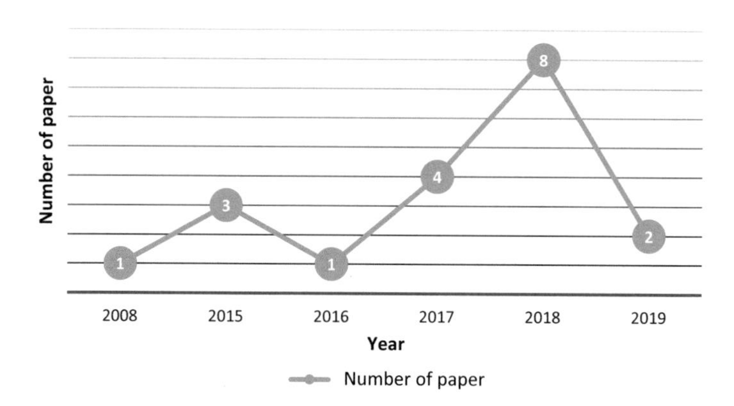


**Fig. 1.2** Publishing trend on Fine–Kinney papers

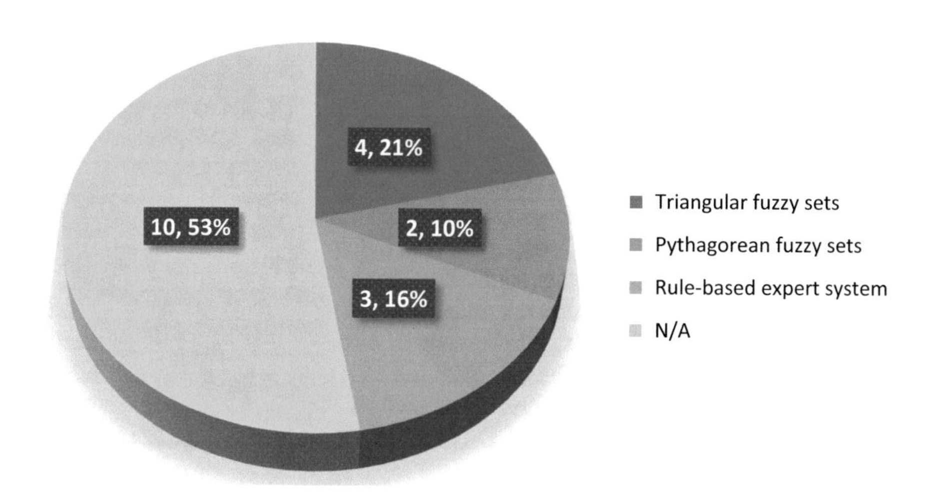


**Fig. 1.3** Distribution of fuzzy set extensions used in Fine–Kinney papers

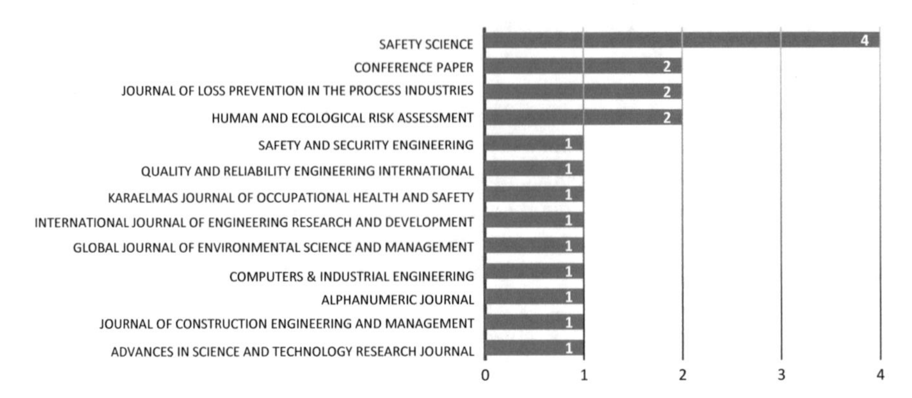


**Fig. 1.4** Distribution of Fine–Kinney papers in terms of journals

that there has been an upward trend recently. From Fig. 1.2, the significant increase in the number of articles after 2016 can be easily seen. The trend has a high $R^2$ value (85%). This suggests that papers about Fine–Kinney are increasing in the occupational health and safety risk assessment literature. This coefficient was determined as a result of a polynomial regression analysis.

Figure 1.3 shows the distribution of Fine–Kinney-based studies using various versions of fuzzy sets. Regarding studies using at least one version of fuzzy sets in related applications, 21% of them (n = 4, in 19 papers) prefer triangular fuzzy numbers. The most preferred fuzzy number versions are as follows: rule-based fuzzy expert system (n = 3, 16%) and Pythagoras fuzzy sets (n = 2, 11%). In the remaining 10 studies, a fuzzy-based model is not used.

Figure 1.4 shows the distribution of the academic journals in which studies have been published. Safety Science has the most publications on Fine–Kinney-based risk assessment approaches (n = 4), followed by Human and Ecological Risk Assessment (n = 2), and Journal of Loss Prevention in the Process Industries (n = 2).

Regarding the application areas, “Manufacturing” constitutes nearly more than a quarter of the total papers. Four papers are concentrated in this application area (Fig. 1.5). They focus on specific problems in maintenance workshop, assembly’s work with welding, aluminum extrusion industry, and gas meter manufacturing plant. Another most studied application areas are “Construction” and “Marine” by four and two papers, respectively. “Transportation”, “Safety”, “Supply chain”, “Textile”, “Energy”, and “Defense” are other application disciplines of Fine–Kinney.

Two important phases of an occupational risk assessment study are comparative analysis and sensitivity analysis. Whether any of these two analyzes are performed is an important factor affecting the quality of the occupational risk analysis. Results of the reviewed papers show that the percentage of studies with comparative analysis is quite high (n = 13, 68%) (Fig. 1.6). On the contrary, the percentage of sensitivity analysis is quite low (n = 1, 5%).

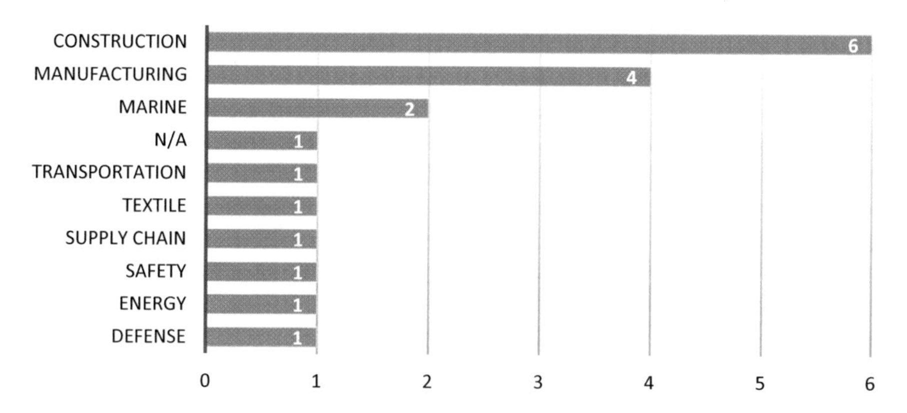


Fig. 1.5 Distribution of Fine–Kinney papers in terms of application area

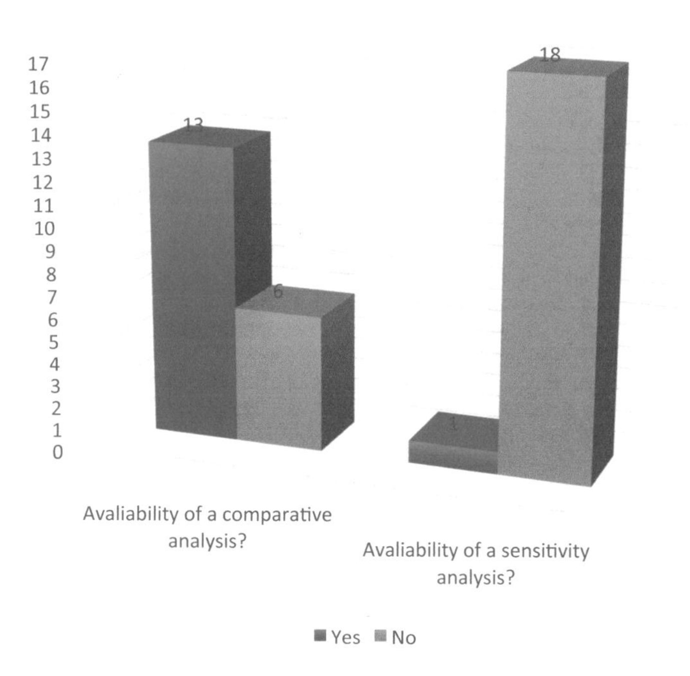


Fig. 1.6 Situation analysis of comparative and sensitivity analysis

## 1.6 Conclusion

In this chapter, the basics of Fine–Kinney method are initially introduced, including its implementing procedure, basic terminology, and drawbacks. Then, a state-of-the-art review of Fine–Kinney occupational risk assessment method and its extensions by fuzzy sets are provided. Graphical results obtained from the review are demonstrated to show the current state. Fuzzy extensions such as triangular fuzzy sets and Pythagorean fuzzy sets are at the forefront of all extensions. They make the classical Fine–Kinney more representative and workable in handling practical and theoretical risk assessment problems. On the other hand, classical Fine–Kinney dealing with hesitant, intuitionistic, interval type-2, and neutrosophic fuzzy-sets-based MCDM approaches are not yet proposed up to now. In the following chapters, considering the drawbacks mentioned in Sect. 1.2, new and novel occupational risk assessment approaches will be presented.

## References


1. Fine, W. T. (1971). Mathematical evaluations for controlling hazards. *Journal of Safety Research, 3*(4), 157–166.
2. Kinney, G. F., & Wiruth, A. D. (1976). *Practical risk analysis for safety management* (pp. 1–20). Naval Weapons Center.
3. Dickson, T. J. (2002). Calculating risks: Fine's mathematical formula 30 years later. *Journal of Outdoor and Environmental Education, 6*(1), 31–39.
4. Birgören, B. (2017). Calculation challenges and solution suggestions for risk factors in the risk analysis method in the Fine Kinney risk analysis method. *Uluslararası Mühendislik Araştırma ve Geliştirme Dergisi, 9*(1), 19–25.
5. Gul, M. (2018). A review of occupational health and safety risk assessment approaches based on multi-criteria decision-making methods and their fuzzy versions. *Human and Ecological Risk Assessment: An International Journal, 24*(7), 1723–1760.
6. Kokangül, A., Polat, U., & Dağsuyu, C. (2017). A new approximation for risk assessment using the AHP and Fine Kinney methodologies. *Safety Science, 91,* 24–32.
7. Karasan, A., Ilbahar, E., Cebi, S., & Kahraman, C. (2018). A new risk assessment approach: Safety and Critical Effect Analysis (SCEA) and its extension with Pythagorean fuzzy sets. *Safety Science, 108,* 173–187.
8. Ilbahar, E., Karaşan, A., Cebi, S., & Kahraman, C. (2018). A novel approach to risk assessment for occupational health and safety using Pythagorean fuzzy AHP & fuzzy inference system. *Safety Science, 103,* 124–136.
9. Gul, M., Celik, E., & Akyuz, E. (2017). A hybrid risk-based approach for maritime applications: The case of ballast tank maintenance. *Human and Ecological Risk Assessment: An International Journal, 23*(6), 1389–1403.
10. Gul, M., Guven, B., & Guneri, A. F. (2018). A new Fine–Kinney-based risk assessment framework using FAHP-FVIKOR incorporation. *Journal of Loss Prevention in the Process Industries, 53,* 3–16.
11. Gul, M., Guneri, A. F., & Baskan, M. (2018). An occupational risk assessment approach for construction and operation period of wind turbines. *Global Journal of Environmental Science and Management, 4*(3), 281–298.
12. Wang, W., Liu, X., & Qin, Y. (2018). A fuzzy Fine–Kinney-based risk evaluation approach with extended MULTIMOORA method based on Choquet integral. *Computers & Industrial Engineering, 125,* 111–123.
13. Oturakçı, M., Dağsuyu, C., & Kokangül, A. (2015). A new approach to Fine Kinney method and an implementation study. *Alphanumeric Journal, 3*(2), 83–92.
14. Marhavilas, P. K., & Koulouriotis, D. E. (2008). A risk-estimation methodological framework using quantitative assessment techniques and real accidents' data: Application in an aluminum extrusion industry. *Journal of Loss Prevention in the Process Industries, 21*(6), 596–603.

15. Gurcanli, G. E., Bilir, S., & Sevim, M. (2015). Activity based risk assessment and safety cost estimation for residential building construction projects. *Safety Science, 80,* 1–12.
16. Netro, Z. G. C., Romero, E. D. L. T., & Flores, J. L. M. (2018). Adaptation of the Fine–Kinney method in supply chain risk assessment. *WIT Transactions on The Built Environment, 174,* 43–55.
17. Korkmaz, E., Iskender, G., & Babuna, F. G. (2016, October). Assessment of occupational health and safety for a gas meter manufacturing plant. In *IOP Conference Series: Earth and Environmental Science* (Vol. 44, No. 3, p. 032015). IOP Publishing.
18. Makajić-Nikolić, D., Kuzmanović, M., & Panić, B. (2018). Terrorism risks assessment of tourism destinations. In *XIII Balkan Conference on Operational Research (BALCOR)* (pp. 341–348). The Mathematical Institute of the Serbian Academy of Sciences and Arts.
19. Birgören, B. (2017). Calculation challenges and solution suggestions for risk factors in the risk analysis method in the Fine Kinney risk analysis method. *International Journal of Engineering Research and Development, 9*(1), 19–25.
20. Gul, M., & Celik, E. (2018). Fuzzy rule-based Fine–Kinney risk assessment approach for rail transportation systems. *Human and Ecological Risk Assessment: An International Journal, 24*(7), 1786–1812.
21. Supciller, A. A., & Abali, N. (2015). Occupational health and safety within the scope of risk analysis with fuzzy proportional risk assessment technique (fuzzy PRAT). *Quality and Reliability Engineering International, 31*(7), 1137–1150.
22. Oturakçı, M., & Dağsuyu, C. (2017). Fuzzy Fine–Kinney approach in risk assessment and an application. *Karaelmas Journal of Occupational Health and Safety, 1*(1), 17–25.
23. Yilmaz, F., & Ozcan, M. S. (2019). A risk analysis and ranking application for lifting vehicles used in construction sites with integrated AHP and Fine–Kinney approach. *Advances in Science and Technology Research Journal, 13*(3), 152–161.
24. Zhang, X., Xing, X., Xie, Y., Zhang, Y., Xing, Z., & Luo, X. (2019). Airport operation situation risk assessment: Combination method based on FAHP and Fine Kinney. *Journal of Construction Engineering and Management.*

# Chapter 2
# Fine–Kinney-Based Occupational Risk Assessment Using Fuzzy Best and Worst Method (F-BWM) and Fuzzy MAIRCA

**Abstract** The best-worst method (BWM) proposed by Rezaei (Omega 53:49–57, [1]) is an MCDM method used to achieve the weights of the criteria by making fewer pairwise comparisons and using more consistent decision matrices. In this chapter, instead of crisp numbers, triangular fuzzy numbers that reflect the uncertainty well in real-world problems are used in integration with the BWM method (Guo and Zhao in Knowl Based Syst 121:23–31, [2]) in determining factor weights. Using the fuzzy best and worst method (F-BWM), a model based on Fine–Kinney occupational health and safety risk assessment method was developed for the first time in the literature. Three parameters of Fine–Kinney method are weighted by the mathematical models of F-BWM. Then, the risks are prioritized by fuzzy multi-attribute ideal real comparative analysis (F-MAIRCA). A case study was conducted to demonstrate the feasibility of the approach, and besides this case study, a comparative study was also conducted to test the validity of the proposed approach. This approach led to the conclusion that Fine–Kinney's method, BWM, MAIRCA, and triangular fuzzy sets make the risk decision-making process more dynamic, taking into account the benefits of these methods individually or in integration.

## 2.1 Fuzzy Set Theory, BWM, and MAIRCA

### *2.1.1 General View on Fuzzy Set Theory*

In this subsection, a general view on the notations regarding fuzzy sets is provided. The fuzzy sets first introduced by Lotfi A. Zadeh in 1965 deal with the uncertainty of human decisions and evaluations in the decision-making process. Real-world decision-making problems involve fuzziness and uncertainty, as decisions, goals, constraints, and possible actions are not fully known [3]. In combining the various experiences, opinions, ideas, and motivations of the individual or group decision-maker, it is better to turn their linguistic terms into fuzzy numbers. Therefore, group decision-making problems practically produced fuzzy numbers. A triangular fuzzy number $\tilde{A}$ can be defined as $\tilde{A}_2 = (l, m, u)$, where *l, m,* and $u$ denote its lower,



M. Gul et al., *Fine–Kinney-Based Fuzzy Multi-criteria Occupational Risk Assessment*, Studies in Fuzziness and Soft Computing 398,
https://doi.org/10.1007/978-3-030-52148-6_2

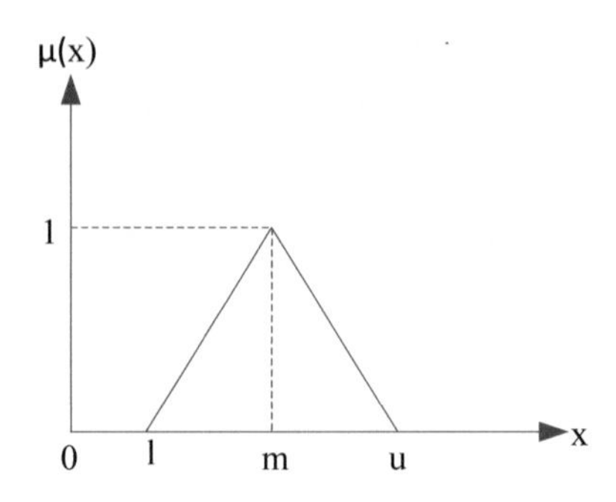


**Fig. 2.1** Triangular fuzzy number

**Table 2.1** The linguistic scale and its corresponding fuzzy numbers

| Linguistic term | Triangular fuzzy number |
|---|---|
| Equal importance (EI) | (1, 1, 1) |
| Weak importance (WI) | (2/3, 1, 1.5) |
| Fair important (FI) | (1.5, 2, 2.5) |
| Very important (VI) | (2.5, 3, 3.5) |
| Absolute importance (AI) | (3.5, 4, 4.5) |

medium, and upper number. Here, an inequality of ($l \leq$ m $\leq$ u) must be satisfied. The membership function of this triangular fuzzy number is as follows (Fig. 2.1):

$$\mu_{\tilde{A}} = \begin{cases} 0, & x < l \\ (x-l)/(m-l), & l \leq x \leq m \\ (u-x)/(u-m), & m \leq x \leq u \\ 0, & x \geq u \end{cases} \tag{2.1}$$

Figure 2.1 also demonstrates a triangular fuzzy number with its membership function. A scale including the linguistic terms and corresponding triangular fuzzy numbers is given in Table 2.1.

For any two triangular fuzzy numbers $\tilde{A}_1 = (l_1, m_1, u_1)$ and $\tilde{A}_2 = (l_2, m_2, u_2)$, the algebraic operations of the two triangular fuzzy numbers can be calculated as follows:

The addition operation;

$$\tilde{A}_1 + \tilde{A}_2 = (l_1 + l_2, m_1 + m_2, u_1 + u_2) \tag{2.2}$$

The subtraction operation;

$$\tilde{A}_1 - \tilde{A}_2 = (l_1 - u_2, m_1 - m_2, u_1 - l_2) \tag{2.3}$$

The multiplication operation;

$$\tilde{A}_1 \times \tilde{A}_2 = (l_1 \times l_2, m_1 \times m_2, u_1 \times u_2) \tag{2.4}$$

The multiplication by a fixed number operation;

$$k \times \tilde{A}_1 = (k \times l_1, k \times m_1, k \times u_1), (k > 0) \tag{2.5}$$

The division by a fixed number operation;

$$\frac{\tilde{A}_1}{k} = \left(\frac{l_1}{k}, \frac{m_1}{k}, \frac{u_1}{k}\right), (k > 0) \tag{2.6}$$

The graded mean integration representation (GMIR) $R\left(\tilde{A}_i\right)$ of a triangular fuzzy number for the ranking is calculated as follows:

$$R\left(\tilde{A}_i\right) = \frac{l_i + 4m_i + u_i}{6}. \tag{2.7}$$

### 2.1.2 BWM Method

The traditional BWM method was initially created by Rezaei [1] as an MCDM method regarding weight assignment to the criteria with fewer pairwise comparison matrices and more consistent matrices.

The main steps of BWM are as follows:

1. Describe the problem and its decision criteria,
2. Define the most crucial and the least crucial criterion,
3. Define the preference of the most crucial criterion over all the other criteria (Best-to-others vector),
4. Define the preference of the least crucial criterion over all the other criteria (Worst-to-others vector),
5. Check the consistency, and
6. Find importance weights of the criteria.

We consider a set of criteria $(e_1, e_2, \ldots, e_n)$ and then select the most crucial criterion and compare with other criteria by Saaty's traditional 9-point scale. To this end, this provides Best-to-others vector as: $E_a = (e_{a1}, e_{a2}, \ldots, e_{an})$, and $e_{aa} = 1$. However, the Worst-to-others vector is: $E_b = (e_{1b}, e_{2b}, \ldots, e_{nb})^T$ by using the same scale.

After deriving the importance weights, the consistency check has been performed using the following formula:

**Table 2.2** Consistency index (CI) values*

| $e_{ab}$ | 1 | 2 | 3 | 4 | 5 | 6 | 7 | 8 | 9 |
|---|---|---|---|---|---|---|---|---|---|
| Consistency Index (max $\xi$) | 0.0 | 0.44 | 1.0 | 1.63 | 2.3 | 3.0 | 3.73 | 4.47 | 5.23 |

$$\xi^2 - (1 + 2u_{BW})\xi + (u_{BW}^2 - u_{BW}) = 0$$

$$\text{Consistency Ratio} = \frac{\xi^*}{\text{Consistency Index}} \tag{2.8}$$

where Table 2.2 shows consistency index values:

To obtain an importance weight value for each criterion, the values of $\left|\frac{w_a}{w_j} - e_{aj}\right|$ and $\left|\frac{w_j}{w_b} - e_{jb}\right|$ for all $j$ must be minimized. To a positive sum for the importance weights, the following mathematical modeling problem must be solved:

$$\begin{aligned}
&\min \max_j \left\{ \left|\frac{w_a}{w_j} - e_{aj}\right|, \left|\frac{w_j}{w_b} - e_{jb}\right| \right\} \\
&s.t. \\
&\sum_j b_j = 1 \\
&b_j \geq 0, \text{ for all } j
\end{aligned} \tag{2.9}$$

The mathematical modeling problem is then transferred to the following one:

$$\begin{aligned}
&\min \xi \\
&s.t. \\
&\left|\frac{w_a}{w_j} - e_{aj}\right| \leq \xi, \text{ for all } j \\
&\left|\frac{w_j}{w_b} - e_{jb}\right| \leq \xi, \text{ for all } j \\
&\sum_j b_j = 1 \\
&b_j \geq 0, \text{ for all } j
\end{aligned} \tag{2.10}$$

By solving this problem, we obtain the optimal importance weights and $\xi^*$.

### 2.1.3 F-MAIRCA Method

MAIRCA is one of the recently developed MCDM methods initially proposed by the Centre for Logistic Research of Defence University of Belgrade by Pamučar et al. [4]. This method has been applied to various problems and application areas like other popular MCDM methods (TOPSIS, PROMETHEE, DEMATEL, ELECTRE, etc.). It uses a simple mathematical algorithm, and is based on the concept of ideal and negative-ideal solutions as in TOPSIS. To incorporate the more imprecision or vagueness, in this chapter, we have injected F-MAIRCA to our proposed risk assessment model. The stages of algorithms are derived from Boral et al. [5] and summarized as follows:

1. Set up the problem and obtain the linguistic initial decision matrices from experts,
2. Construct a fuzzy aggregated decision matrix,
3. Determine the preferences of alternatives,
4. Determine the matrix of fuzzy theoretical ponder,
5. Construct the normalized initial fuzzy aggregated decision matrix,
6. Determine the matrix of fuzzy actual ponder,
7. Compute the distance between the matrix of fuzzy theoretical ponder and the matrix of fuzzy actual ponder,
8. Sum the gap values, and
9. Rank the alternatives.

## 2.2 Proposed Fine–Kinney-Based Approach Using F-BWM and F-MAIRCA

In F-BWM, there are $n$ criteria. The fuzzy pairwise comparisons regarding $n$ criteria are constructed using the linguistic scale provided in Table 2.1. Then, the linguistic terms evaluated by decision-makers are transformed into triangular fuzzy numbers. Thus, the fuzzy decision matrix (pairwise comparison matrix) is obtained as follows:

$$\tilde{A} = \begin{array}{c} \\ c_1 \\ c_2 \\ \vdots \\ c_n \end{array} \begin{array}{c} \begin{array}{cccc} c_1 & c_2 & \cdots & c_n \end{array} \\ \begin{bmatrix} \tilde{a}_{11} & \tilde{a}_{12} & \cdots & \tilde{a}_{1n} \\ \tilde{a}_{21} & \tilde{a}_{22} & \cdots & \tilde{a}_{2n} \\ \vdots & \vdots & \ddots & \vdots \\ \tilde{a}_{n1} & \tilde{a}_{n2} & \cdots & \tilde{a}_{nn} \end{bmatrix} \end{array}$$

where $\tilde{a}_{ij}$ refers to the relative fuzzy preference of criterion $i$ to criterion $j$, which is a triangular fuzzy number; $\tilde{a}_{ij} = (1, 1, 1)$ when $i = j$.

In this chapter, we will base the original study [2] to present the detailed steps of F-BWM.

**Step 1: Criteria set design**. Here in this research, there are $n$ decision criteria $\{c_1, c_2, \ldots, c_n\}$.

**Step 2: Identify the best and the worst criterion**. Here, the best criterion and the worst criterion are indicated as $c_B$ and $c_W$, respectively.

**Step 3: Do the fuzzy reference comparisons for the best criterion**. The fuzzy Best-to-Others vector is: $\tilde{A}_B = \{\tilde{a}_{B1}, \tilde{a}_{B2}, \ldots, \tilde{a}_{Bn}\}$.

where $\tilde{a}_{Bj}$ refers to the fuzzy preference of the best criterion $c_B$ over criterion $j$, $j = 1, 2, \ldots, n$. It is a common rule that $\tilde{a}_{BB} = (1, 1, 1)$.

**Step 4: Do the fuzzy reference comparisons for the worst criterion**. The fuzzy Others-to-Worst vector can be obtained as: $\tilde{A}_W = \{\tilde{a}_{1W}, \tilde{a}_{2W}, \ldots, \tilde{a}_{nW}\}$.

where $\tilde{a}_{iW}$ refers to the fuzzy preference of criterion $i$ over the worst criterion $c_W$, $i = 1, 2, \ldots, n$.

**Step 5: Check the consistency**. The consistency ratio is determined in the same way of BWM.

**Step 6: Find the optimal importance weights** $\left(\tilde{w}_1^*, \tilde{w}_2^*, \ldots, \tilde{w}_n^*\right)$. The same inferences as in crisp BWM are used to reach the optimal weights. It should be noted that $\tilde{w}_B$, $\tilde{w}_j$, and $\tilde{w}_W$ in F-BWM are triangular fuzzy numbers, which are different from BWM. In some cases, we prefer to use $\tilde{w}_j = \left(l_j^w, m_j^w, u_j^w\right)$ for an optimal alternative selection. The fuzzy weight of criterion denoted by triangular fuzzy numbers $\tilde{w}_j = \left(l_j^w, m_j^w, u_j^w\right)$ needs to be transformed to a crisp value. In this chapter, we used Eq. 2.7 for transforming triangular fuzzy numbers to crisp numbers. Consequently, we can determine the constrained optimization problem for obtaining the optimal fuzzy weights $\left(\tilde{w}_1^*, \tilde{w}_2^*, \ldots, \tilde{w}_n^*\right)$ as follows.

$$\min \max_j \left\{ \left| \frac{\tilde{w}_B}{\tilde{w}_j} - \tilde{a}_{Bj} \right|, \left| \frac{\tilde{w}_j}{\tilde{w}_W} - \tilde{a}_{jW} \right| \right\}$$

$$s.t. \begin{cases} \sum_{j=1}^{n} R(\tilde{w}_i) = 1 \\ l_j^w \le m_j^w \le u_j^w \\ l_j^w \ge 0 \\ j = 1, 2, \ldots, n \end{cases} \tag{2.11}$$

where $\tilde{w}_B = \left(l_B^w, m_B^w, u_B^w\right)$, $\tilde{w}_j = \left(l_j^w, m_j^w, u_j^w\right)$, $\tilde{w}_W = \left(l_W^w, m_W^w, u_W^w\right)$, $\tilde{a}_{Bj} = \left(l_{Bj}^w, m_{Bj}^w, u_{Bj}^w\right)$, and $\tilde{a}_{jW} = \left(l_{jW}^w, m_{jW}^w, u_{jW}^w\right)$. Equation (2.11) can be transferred to the following problem.

$$\min \xi$$
$$s.t.\begin{cases} \left|\frac{\tilde{w}_B}{\tilde{w}_j} - \tilde{a}_{Bj}\right| \le \xi \\ \left|\frac{\tilde{w}_j}{\tilde{w}_W} - \tilde{a}_{jW}\right| \le \xi \\ \sum_{j=1}^{n} R(\tilde{w}_i) = 1 \\ l_j^w \le m_j^w \le u_j^w \\ l_j^w \ge 0 \\ j = 1, 2, \ldots, n \end{cases} \tag{2.12}$$

where $\xi = \left(l^\xi, m^\xi, u^\xi\right)$

Considering $l^\xi \le m^\xi \le u^\xi$, $\xi^* = (k^*, k^*, k^*)$, $k^* \le l^\xi$ then Eq. (2.12) can be transferred as

$$\min \xi^*$$
$$s.t.\begin{cases} \left|\frac{\left(l_B^w, m_B^w, u_B^w\right)}{\left(l_j^w, m_j^w, u_j^w\right)} - \left(l_{Bj}, m_{Bj}, u_{Bj}\right)\right| \le \left(k^*, k^*, k^*\right) \\ \left|\frac{\left(l_j^w, m_j^w, u_j^w\right)}{\left(l_W^w, m_W^w, u_W^w\right)} - \left(l_{jW}, m_{jW}, u_{jW}\right)\right| \le \left(k^*, k^*, k^*\right) \\ \sum_{j=1}^{n} R(\tilde{w}_i) = 1 \\ l_j^w \le m_j^w \le u_j^w \\ l_j^w \ge 0 \\ j = 1, 2, \ldots, n \end{cases} \tag{2.13}$$

**Step 7: Assess the risks by F-MAIRCA method considering weighted Fine–Kinney parameters by F-BWM**. In this step, F-MAIRCA method can be applied to the solution of the problem. The focal point here is to show the applicability of F-BWM incorporated with Fine–Kinney occupational risk assessment which is the core point of the book.

The suggested approach is demonstrated as a step-by-step procedure in Fig. 2.2.

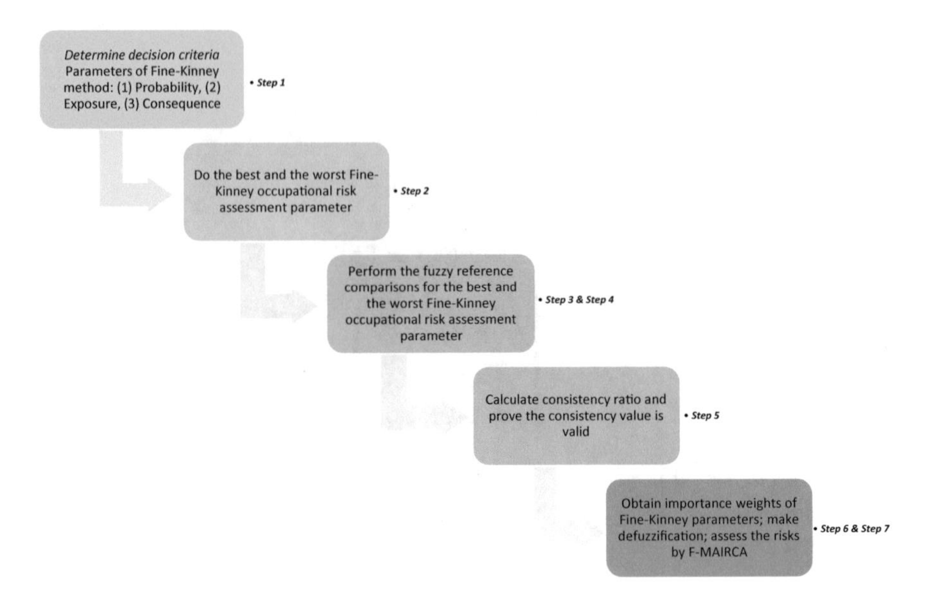


**Fig. 2.2** The main steps used in the suggested approach

## 2.3 Case Study

To demonstrate the applicability of the proposed approach, a real case study is carried out for the occupational risk assessment of raw mill processes in a cement factory. As the initial step of the approach, the F-BWM method was applied to compute the weight values of Fine–Kinney risk assessment parameters. Furthermore, the reliability of the pairwise comparisons consistency in F-BWM was checked, and it was found acceptable for the matrix. In the second stage, F-MAIRCA was used to rank hazards and their associated risks. In the following, step-by-step application of the proposed approach to the problem is provided.

### *2.3.1 Application Results*

In this application, three parameters such as probability ($P$), exposure ($E$), and consequence ($C$) are handled. Also, the risk list regarding the raw mill processes in the observed cement factory is provided in Table 2.3. The consequence ($C$) and exposure ($E$) are determined as the best and the worst parameter, respectively, (Step 2). The fuzzy reference comparisons are applied, and the linguistic terms for fuzzy preferences of the best parameter and the worst parameter are given in Table 2.4 and Table 2.5, respectively.

**Table 2.3** The hazard identification list

| Hazard ID | Activity area | Hazard identification |
|---|---|---|
| Hazard-1 | Coming near and moving away of the Reclaimer car from the stack | Rotary moving parts of machinery and its equipment |
| Hazard-2 | Cleaning of the bucket fronts near the Reclaimer shot | Rotary moving parts of machinery and its equipment |
| Hazard-3 | | Slippery floor, inadequate and untidy working area |
| Hazard-4 | | Non-ergonomic equipment/movement |
| Hazard-5 | Cleaning the Reclaimer shot | Fall from height |
| Hazard-6 | | Non-ergonomic equipment/movement |
| Hazard-7 | | Slippery floor, inadequate and untidy working area |
| Hazard-8 | | Manual handling, lifting, installation, loading, forcing |
| Hazard-9 | Cleaning of band relays and band bottom | Rotary moving parts of machinery and its equipment |
| Hazard-10 | | Slippery floor, inadequate and untidy working area |
| Hazard-11 | | Non-ergonomic equipment/movement |
| Hazard-12 | | Fall from height |
| Hazard-13 | | Stationary and mobile ladders and platforms |
| Hazard-14 | | Unannounced start-up |

**Table 2.4** The evaluations in linguistic variables from the best parameter side

| Fine–Kinney risk parameter | P | E | C |
|---|---|---|---|
| Best factor (C) | FI | VI | EI |

**Table 2.5** The evaluations in linguistic variables from the worst parameter side

| Fine–Kinney risk parameter | Worst factor (E) |
|---|---|
| P | WI |
| E | EI |
| C | VI |

Then, the fuzzy Best-to-Others vector and the fuzzy Others-to-Worst can be obtained with respect to Table 2.1 as follows (Step 3).

$$\tilde{a}_B = [(3/2,\ 2,\ 5/2), (5/2,\ 3,\ 7/2), (1,\ 1,\ 1)]$$

$$\tilde{a}_W = [(2/3,\ 1,\ 3/2)\ (1,\ 1,\ 1), (5/2,\ 3,\ 7/2)]$$

Then, for obtaining the optimal fuzzy weights of three parameters, the nonlinearly constrained model is constructed as follows:

$$\min \xi^*$$

$$s.t.\begin{cases} \left|\dfrac{(l_P^w, m_P^w, u_P^w)}{(l_P^w, m_P^w, u_P^w)} - (l_{PP}, m_{PP}, u_{PP})\right| \leq (e^*, e^*, e^*) \\ \left|\dfrac{(l_P^w, m_P^w, u_P^w)}{(l_E^w, m_E^w, u_E^w)} - (l_{PE}, m_{PE}, u_{PE})\right| \leq (e^*, e^*, e^*) \\ \left|\dfrac{(l_P^w, m_P^w, u_P^w)}{(l_C^w, m_C^w, u_C^w)} - (l_{PC}, m_{PC}, u_{PC})\right| \leq (e^*, e^*, e^*) \\ \left|\dfrac{(l_P^w, m_P^w, u_P^w)}{(l_E^w, m_E^w, u_E^w)} - (l_{PE}, m_{PE}, u_{PE})\right| \leq (e^*, e^*, e^*) \\ \left|\dfrac{(l_E^w, m_E^w, u_E^w)}{(l_E^w, m_E^w, u_E^w)} - (l_{EE}, m_{EE}, u_{EE})\right| \leq (e^*, e^*, e^*) \\ \left|\dfrac{(l_C^w, m_C^w, u_C^w)}{(l_E^w, m_E^w, u_E^w)} - (l_{CE}, m_{CE}, u_{CE})\right| \leq (e^*, e^*, e^*) \\ \sum_{j=P,E,C} R(\tilde{w}_j) = 1 \\ l_j^w \leq m_j^w \leq u_j^w \\ l_j^w \geq 0 \\ j = P, E, C \end{cases}$$

Then, the following nonlinearly constrained optimization problem is obtained using represented by crisp numbers.

$$\min e$$

$$s.t.\begin{cases} l_P - 0.6667 * u_E - u_E * e \leq 0;\ l_P - 0.6667 * u_E + u_E * e \geq 0;\ m_P - 1 * m_E - m_E * e \leq 0; \\ m_P - 1 * m_E + m_E * e \geq 0;\ u_P - 1.5 * l_E - l_E * e \leq 0;\ u_P - 1.5 * l_E + l_E * e \geq 0; \\ l_C - 1.5 * u_P - u_P * e \leq 0;\ l_C - 1.5 * u_P + u_P * e \geq 0;\ m_C - 2 * m_P - m_P * e \leq 0; \\ m_C - 2 * m_P + m_P * e \geq 0;\ u_C - 2.5 * l_P - l_P * e \leq 0;\ u_C - 2.5 * l_P + l_P * e \geq 0; \\ l_C - 2.5 * u_E - u_E * e \leq 0;\ l_C - 2.5 * u_E + u_E * e \geq 0;\ m_C - 3 * m_E - m_E * e \leq 0; \\ m_C - 3 * m_E + m_E * e \geq 0;\ u_C - 3.5 * l_E - l_E * e \leq 0;\ u_C - 3.5 * l_E + l_E * e \geq 0; \\ l_P \leq m_P \leq u_P;\ l_E \leq m_E \leq u_E;\ l_C \leq m_C \leq u_C; \\ \frac{1}{6} * (l_P + 4 * m_P + u_P) + \frac{1}{6} * (l_E + 4 * m_E + u_E) + \frac{1}{6} * (l_C + 4 * m_C + u_C) = 1; \\ l_P > 0;\ l_E > 0;\ l_C > 0 \\ e \geq 0 \end{cases}$$

The optimal fuzzy weights of three parameters (“P”, ”E”, and “C”) are calculated as follows:

$w_p^* = (0.210, 0.246, 0.304)$, $w_E^* = (0.176, 0.199, 0.233)$, $w_C^* = (0.528, 0.550, 0.573)$, and $\xi^* = (0.236, 0.236, 0.236)$.

Then, the crisp weights of these parameters are determined as follows:

$w_p^* = 0.250,\ w_E^* = 0.200,\ w_C^* = 0.550.$

In this process, the consistency ratio is calculated. $\tilde{a}_{Bw} = a_{12} = (5/2, 3, 7/2)$ is the largest in the interval, hence, CI is considered as 6.69 using Table 2.6. The consistency ratio is $CR = 0.236/6.69 = 0.0352$, which shows a good consistency (it is so close to zero).

The F-BWM results showed that "Consequence parameter" (C) is the highest rank followed by "Probability parameter" (P) and "Exposure parameter" (E) as shown in Table 2.7.

In the last step of the proposed approach, prioritizing of the hazards by F-MAIRCA model is performed. Using the linguistic terms given in Table 2.8, we first obtain the fuzzy aggregated decision matrix as shown in Table 2.9.

By following the steps as indicated in Sect. 2.1.3, we obtain the final values of F-MAIRCA model. Using the model, the gap values were directly calculated and hazards are ranked from the ascending order view (Table 2.10). Results of the study demonstrate that the most important three risks are rotary moving parts of machinery and its equipment in the process of Cleaning of the bucket fronts near the Reclaimer

**Table 2.6** Consistency index for F-BWM

| Linguistic term | EI | WI | FI | VI | AI |
|---|---|---|---|---|---|
| $\tilde{a}_{Bw}$ | (1, 1, 1) | (2/3, 1, 3/2) | (3/2, 2, 5/2) | (5/2, 3, 7/2) | (7/2, 4, 9/2) |
| CI | 3 | 3.8 | 5.29 | 6.69 | 8.04 |

**Table 2.7** The obtained importance weight of Fine–Kinney risk parameters

| Factor | Fuzzy weight vector | Defuzzified weight vector | Rank |
|---|---|---|---|
| P | (0.210, 0.246, 0.304) | 0.250 | 2 |
| E | (0.176, 0.199, 0.233) | 0.200 | 3 |
| C | (0.528, 0.550, 0.573) | 0.550 | 1 |

**Table 2.8** The linguistic terms and related fuzzy numbers used in assessing risks

| Linguistic terms | Fuzzy score |
|---|---|
| Very poor (VP) | (0, 0, 1) |
| Poor (P) | (0, 1, 3) |
| Medium poor (MP) | (1, 3, 5) |
| Fair (F) | (3, 5, 7) |
| Medium good (MG) | (5, 7, 9) |
| Good (G) | (7, 9, 10) |
| Very good (VG) | (9, 10, 10) |

**Table 2.9** The initial linguistic values from the OHS experts' consensus

| Hazard ID | Probability | Exposure | Consequence |
|---|---|---|---|
| Hazard-1 | P | P | G |
| Hazard-2 | MP | G | G |
| Hazard-3 | MP | MP | G |
| Hazard-4 | F | G | VP |
| Hazard-5 | F | MP | F |
| Hazard-6 | MG | G | VP |
| Hazard-7 | MP | G | F |
| Hazard-8 | F | G | P |
| Hazard-9 | MP | VP | G |
| Hazard-10 | F | G | MP |
| Hazard-11 | MG | MP | MP |
| Hazard-12 | MP | MP | P |
| Hazard-13 | F | MP | MP |
| Hazard-14 | MP | G | G |

**Table 2.10** Final risk scores and rankings by the F-MAIRCA

| Hazard | Gap values of risk parameters | | | Sum of gap values | Rank |
|---|---|---|---|---|---|
| | P | E | C | | |
| Hazard-1 | 0.017 | 0.014 | 0.030 | 0.0616 | 4 |
| Hazard-2 | 0.016 | 0.012 | 0.030 | 0.0581 | 1 |
| Hazard-3 | 0.016 | 0.013 | 0.030 | 0.0600 | 2 |
| Hazard-4 | 0.015 | 0.012 | 0.039 | 0.0655 | 12 |
| Hazard-5 | 0.015 | 0.013 | 0.034 | 0.0626 | 6 |
| Hazard-6 | 0.014 | 0.012 | 0.039 | 0.0643 | 9 |
| Hazard-7 | 0.016 | 0.012 | 0.034 | 0.0618 | 5 |
| Hazard-8 | 0.015 | 0.012 | 0.038 | 0.0645 | 10 |
| Hazard-9 | 0.016 | 0.014 | 0.030 | 0.0610 | 3 |
| Hazard-10 | 0.015 | 0.012 | 0.036 | 0.0627 | 7 |
| Hazard-11 | 0.014 | 0.013 | 0.036 | 0.0635 | 8 |
| Hazard-12 | 0.016 | 0.013 | 0.038 | 0.0676 | 13 |
| Hazard-13 | 0.015 | 0.013 | 0.036 | 0.0646 | 11 |
| Hazard-14 | 0.016 | 0.012 | 0.030 | 0.0581 | 1 |

shot (Hazard-2), unannounced start-up in the process of cleaning of band relays and band bottom (Hazard-14) and slippery floor, inadequate and untidy working area during the cleaning of the bucket fronts near the Reclaimer shot (Hazard-3).

### 2.3.2 Validation Study on the Results

In this subsection, some validation tests of the obtained ranking results are provided. As a first validation study, we made a comparative study between the results of the existed approach (F-BWM and F-MAIRCA under Fine–Kinney's method) and other popular MCDM-based approaches. We then observe the variations in hazard rankings. The results are shown in Table 2.11.

According to the results obtained from Table 2.11, the results of all solved approaches also conclude that Hazard-2 is the most critical hazard. The sequence then continues with Hazard-14 and Hazard-3. By using the classical Fine–Kinney in a single method setting, Hazard-11 is ranked as the fourth critical hazard apart from the other three approaches. Otherwise, according to study B and study C, Hazard-9 is ranked as the fourth critical hazard. When we compare the results obtained by these approaches, it is observed that a very small rank variation exists between them.

**Table 2.11** First validation results: Comparison of rankings by different approaches

| Hazard | Ranking order of the hazard | | | |
|---|---|---|---|---|
| | Study A | Study B | Study C | Study D |
| Hazard-1 | 4 | 6 | 5 | 11 |
| Hazard-2 | 1 | 1 | 1 | 1 |
| Hazard-3 | 2 | 3 | 3 | 3 |
| Hazard-4 | 12 | 13 | 13 | 12 |
| Hazard-5 | 6 | 7 | 8 | 5 |
| Hazard-6 | 9 | 12 | 11 | 8 |
| Hazard-7 | 5 | 5 | 6 | 6 |
| Hazard-8 | 10 | 11 | 10 | 13 |
| Hazard-9 | 3 | 4 | 4 | 7 |
| Hazard-10 | 7 | 8 | 7 | 9 |
| Hazard-11 | 8 | 9 | 9 | 4 |
| Hazard-12 | 13 | 14 | 14 | 14 |
| Hazard-13 | 11 | 10 | 12 | 10 |
| Hazard-14 | 1 | 2 | 2 | 2 |

*Study A* Current approach (F-BWM and F-MAIRCA under Fine–Kinney); *Study B* F-BWM and F-VIKOR [6] under Fine–Kinney; *Study C* F-BWM and F-TOPSIS [7] under Fine–Kinney; *Study D* Classical Fine–Kinney

The highest difference is observed between the classical Fine–Kinney method and others. Although an extreme rank variation between the benchmarking models is not observed between the benchmarking models that have been previously proved in the literature and our current approach, this integrated approach is new in the Fine–Kinney domain.

As a second validation study, we made a sensitivity analysis and analyze the variation in hazard ranking with respect to the changes in parameters' weights. Hence, we interchange the weight vectors of the Fine–Kinney parameters. As there are three parameters in Fine–Kinney, in our case study, a total of six combinations are created. The weight vectors of Fine–Kinney parameters designed for the sensitivity analysis are given in Table 2.12.

**Table 2.12** The weight vectors designed for the sensitivity analysis

| ID | Weight vector | Parameter | Fuzzy weight value | | |
|---|---|---|---|---|---|
| $W_{PEC}$ | Weight vector-1 | Probability | 0.210 | 0.246 | 0.304 |
| | | Exposure | 0.176 | 0.199 | 0.233 |
| | | Consequence | 0.528 | 0.550 | 0.573 |
| $W_{PCE}$ | Weight vector-2 | Probability | 0.210 | 0.246 | 0.304 |
| | | Exposure | 0.528 | 0.550 | 0.573 |
| | | Consequence | 0.176 | 0.199 | 0.233 |
| $W_{EPC}$ | Weight vector-3 | Probability | 0.176 | 0.199 | 0.233 |
| | | Exposure | 0.210 | 0.246 | 0.304 |
| | | Consequence | 0.528 | 0.550 | 0.573 |
| $W_{ECP}$ | Weight vector-4 | Probability | 0.176 | 0.199 | 0.233 |
| | | Exposure | 0.528 | 0.550 | 0.573 |
| | | Consequence | 0.210 | 0.246 | 0.304 |
| $W_{CPE}$ | Weight vector-5 | Probability | 0.528 | 0.550 | 0.573 |
| | | Exposure | 0.210 | 0.246 | 0.304 |
| | | Consequence | 0.176 | 0.199 | 0.233 |
| $W_{CEP}$ | Weight vector-6 | Probability | 0.528 | 0.550 | 0.573 |
| | | Exposure | 0.176 | 0.199 | 0.233 |
| | | Consequence | 0.210 | 0.246 | 0.304 |

The ranking orders with respect to the sensitivity analysis are shown in Table 2.13.

It can be observed from Table 2.13 that when the weight vector varies, there are variations in the ranking orders of hazards. Hence, our method is sensitive to Fine–Kinney risk parameters' weights. Hazard-2 and Hazard-14 are mostly ranked as the most critical two hazards. Their ranking orders have changed according to the last weight vector ($W_{CEP}$). There is no change in the ranking of Hazard-12 and Hazard-13 for all combinations excluding weight vector-5 and 6. When compared to the results with the ones similar to this study from the literature, we can say that the ranking result obtained by our proposed approach is credible and applicable.

**Table 2.13** Ranking order changes in times of parameters' weights changes

| Hazard | Ranking order of the hazard | | | | | |
|---|---|---|---|---|---|---|
| | $W_{PEC}$ | $W_{PCE}$ | $W_{EPC}$ | $W_{ECP}$ | $W_{CPE}$ | $W_{CEP}$ |
| Hazard-1 | 4 | 12 | 5 | 11 | 11 | 13 |
| Hazard-2 | 1 | 1 | 1 | 1 | 1 | 3 |
| Hazard-3 | 2 | 8 | 2 | 7 | 7 | 10 |
| Hazard-4 | 12 | 6 | 12 | 6 | 6 | 8 |
| Hazard-5 | 6 | 9 | 7 | 8 | 8 | 7 |
| Hazard-6 | 9 | 2 | 10 | 4 | 4 | 1 |
| Hazard-7 | 5 | 4 | 4 | 2 | 2 | 9 |
| Hazard-8 | 10 | 5 | 9 | 5 | 5 | 6 |
| Hazard-9 | 3 | 11 | 3 | 12 | 12 | 12 |
| Hazard-10 | 7 | 3 | 6 | 3 | 3 | 5 |
| Hazard-11 | 8 | 7 | 8 | 9 | 9 | 2 |
| Hazard-12 | 13 | 13 | 13 | 13 | 13 | 14 |
| Hazard-13 | 11 | 10 | 11 | 10 | 10 | 11 |
| Hazard-14 | 1 | 1 | 1 | 1 | 1 | 4 |

## 2.4 Python Implementation of the Proposed Approach

```
# Chapter 2
# import required libraries
import numpy as np

# set initial variables
n_criteria = 3  # number of criteria
n_hazard = 14  # number of hazards
n_BWM = 3  # number of BWM values
n_fuzzy = 3  # number of fuzzy values
p_A = 1 / n_hazard
idm = []  # initial decision matrix
ndm = []  # normalized decision matrix
F_BWM = [  # L,    M,   U
    [0.21, 0.246, 0.304],  # P
    [0.176, 0.199, 0.233],  # E
    [0.528, 0.55, 0.573]  # C
]
fuzzy_scores = [
    ["VP", 0, 0, 1],  # Very poor (VP)
    ["P", 0, 1, 3],  # Poor (P)
    ["MP", 1, 3, 5],  # Medium poor (MP)
    ["F", 3, 5, 7],  # Fair (F)
    ["MG", 5, 7, 9],  # Medium good (MG)
    ["G", 7, 9, 10],  # Good (G)
    ["VG", 9, 10, 10]  # Very good (VG)
]

# ehe: 1st expert's hazard evaluation
ehe = [
    # Hazard ID, Probability, Exposure,    Consequence
    ["Hazard-1", "P", "P", "G"],
    ["Hazard-2", "MP", "G", "G"],
    ["Hazard-3", "MP", "MP", "G"],
```

```
    ["Hazard-4", "F", "G", "VP"],
    ["Hazard-5", "F", "MP", "F"],
    ["Hazard-6", "MG", "G", "VP"],
    ["Hazard-7", "MP", "G", "F"],
    ["Hazard-8", "F", "G", "P"],
    ["Hazard-9", "MP", "VP", "G"],
    ["Hazard-10", "F", "G", "MP"],
    ["Hazard-11", "MG", "MP", "MP"],
    ["Hazard-12", "MP", "MP", "P"],
    ["Hazard-13", "F", "MP", "MP"],
    ["Hazard-14", "MP", "G", "G"]
]

def rank(vector, da):  # da -1:descending, 1:ascending
    order = np.zeros([len(vector), 1])
    unique_val = da * np.sort(da * np.unique(vector))
    for ix in range(0, len(unique_val)):
        order[np.argwhere(vector == unique_val[ix])] = ix + 1
    return order

def print_result(order, vector):
    print('Hazard Id, Rank, Value')
    for ix in range(0, len(order)):
        print(ehe[ix][0], ', ', int(order[ix]), ', ', vector[ix])

# set initial decision matrix
for hz in ehe:
    temp_var = []  # temporary variable
    for cr in range(0, n_criteria):
        for fs in fuzzy_scores:
            if hz[cr + 1] == fs[0]:
                temp_var.append(fs[1:])
    idm.append(temp_var)

# set initial aggregate matrix (iam)
iam = np.sum(np.square(idm), axis=2)

# set normalized decision matrix
temp_var = np.sqrt(np.sum(iam, axis=0))

for rw in idm:
    temp_var2 = []
    for j in range(0, n_criteria):
        temp_var2.append(rw[j] / temp_var[j])
    ndm.append(temp_var2)

# set q value
tp = np.multiply(F_BWM, p_A)  # tp: theoretical_ponder
ap = ndm * tp  # ap: actual_ponder
sum_of_gap = np.sum(np.sqrt(np.sum(np.square(tp - ap), axis=2) / 3), axis=1)
```

```
hazard_rank = rank(sum_of_gap, 1)
print_result(hazard_rank, sum_of_gap)

'''
Output:
Hazard Id, Rank, Value
Hazard-1 , 4 , 0.061623349450804804
Hazard-2 , 1 , 0.05806150565291705
Hazard-3 , 2 , 0.05997926137974485
Hazard-4 , 12 , 0.0655328467463496
Hazard-5 , 6 , 0.06257321299623751
Hazard-6 , 9 , 0.06434445478286971
Hazard-7 , 5 , 0.06184422379555565
Hazard-8 , 10 , 0.06446020833920928
Hazard-9 , 3 , 0.06097932279472583
Hazard-10 , 7 , 0.0627238536398366
Hazard-11 , 8 , 0.06345321740318449
Hazard-12 , 13 , 0.06756673059218302
Hazard-13 , 11 , 0.06464160936666438
Hazard-14 , 1 , 0.05806150565291705
'''
```

## References


1. Rezaei, J. (2015). Best-worst multi-criteria decision-making method. *Omega, 53,* 49–57.
2. Guo, S., & Zhao, H. (2017). Fuzzy best-worst multi-criteria decision-making method and its applications. *Knowledge-Based Systems, 121,* 23–31.
3. Zadeh, L. A. (1965). Fuzzy sets. *Information and Control, 8*(3), 338–353.
4. Pamučar, D., Vasin, L., & Lukovac, L. (2014, October). Selection of railway level crossings for investing in security equipment using hybrid DEMATEL-MARICA model. In *XVI International Scientific-expert Conference on Railway, Railcon* (pp. 89–92).
5. Boral, S., Howard, I., Chaturvedi, S. K., McKee, K., & Naikan, V. N. A. (2019). An integrated approach for fuzzy failure modes and effects analysis using fuzzy AHP and fuzzy MAIRCA. *Engineering Failure Analysis*, 104195.
6. Gul, M. (2018). Application of Pythagorean fuzzy AHP and VIKOR methods in occupational health and safety risk assessment: The case of a gun and rifle barrel external surface oxidation and colouring unit. *International Journal of Occupational Safety and Ergonomics*, 1–14.
7. Gul, M., & Guneri, A. F. (2016). A fuzzy multi criteria risk assessment based on decision matrix technique: A case study for aluminum industry. *Journal of Loss Prevention in the Process Industries, 40,* 89–100.

# Chapter 3
# Fine–Kinney-Based Occupational Risk Assessment Using Interval Type-2 Fuzzy TOPSIS

**Abstract** This chapter proposed an improved Fine–Kinney occupational risk assessment approach using a well-known MCDM method "TOPSIS" under interval type-2 fuzzy set concept. It is defined as a technique for order preference by similarity to ideal solution by Hwang and Yoon [1]. It is based on separation from ideal and anti-ideal solution concept. Since the initial crisp data-based version is insufficient by time in reflecting the uncertainty in decision-maker's opinions, fuzzy sets are integrated to the TOPSIS algorithm to provide a solid and comprehensive method. Interval type-2 fuzzy set is an improved version of the type-1 fuzzy set. It is also a special version of a general type-2 fuzzy set. Since general type-2 fuzzy systems contain complex computational operations, they cannot be easily applied to real-world problems such as occupational risk assessment. Interval type-2 fuzzy sets are the most frequently used type-2 fuzzy sets due to their ability in handling more uncertainty and producing more accurate and solid results. The Fine–Kinney concept is merged with the interval type-2 fuzzy set concept and TOPSIS for the first time through the literature. To demonstrate the applicability of the proposed approach, a case study is carried out in a chrome plating unit of a gun factory. Some beneficial validation and sensitivity analysis are also performed. Finally, as a creative contribution of our book, the implementation of the proposed approach in Python is performed.

## 3.1 Interval Type-2 Fuzzy Set and TOPSIS

### *3.1.1 General View on Interval Type-2 Fuzzy Set*

Prior to defining the proposed interval type-2 F-TOPSIS method, a general view on the definitions of type-2 fuzzy sets is provided with the aid of some important reference studies performed by Mendel et al. [2], Lee and Chen [3], Chen and Lee [4], Celik et al. [5–8], Soner et al. [9], Demirel et al. [10], Celik and Gumus [11, 12], Kahraman et al. [13] and Celik [14]:



M. Gul et al., *Fine–Kinney-Based Fuzzy Multi-criteria Occupational Risk Assessment*, Studies in Fuzziness and Soft Computing 398,
https://doi.org/10.1007/978-3-030-52148-6_3

i. A type-2 fuzzy set $\tilde{\tilde{A}}$ with its membership function $\mu_{\tilde{\tilde{A}}}$, can be demonstrated as follows:

$$\tilde{\tilde{A}} = \left\{((x, u), \mu_{\tilde{\tilde{A}}}(x, u))|\forall x \in X, \forall u \in J_X \subseteq [0, 1], 0 \leq \mu_{\tilde{\tilde{A}}}(x, u) \leq 1\right\}$$

where $J_X$ denotes an interval in [0, 1].

ii. If all $\mu_{\tilde{\tilde{A}}}(x, u) = 1$, then $\tilde{\tilde{A}}$ is named as an interval type-2 fuzzy set. It is considered a special branch in the type-2 fuzzy set.

iii. The upper and the lower membership functions in this special type fuzzy set are the same with type-1 membership functions. The reference points and the heights of the membership functions are arranged according to the study of Chen and Lee [4]. Figure 3.1 shows a graphical view of a trapezoidal interval type-2 fuzzy set. It is also formulized as follows:

$$\tilde{\tilde{A}}_i = \left(\tilde{A}_i^U, \tilde{A}_i^L\right) = \left(\left(a_{i1}^U, a_{i2}^U, a_{i3}^U, a_{i4}^U; H_1\left(\tilde{A}_i^U\right), H_2\left(\tilde{A}_i^U\right)\right), \left(a_{i1}^L, a_{i2}^L, a_{i3}^L, a_{i4}^L; H_1\left(\tilde{A}_i^L\right), H_2\left(\tilde{A}_i^L\right)\right)\right)$$

Here the meaning of the notations are as follows:

- $\tilde{A}_i^U, \tilde{A}_i^L$: Type-1 fuzzy sets,
- $a_{i1}^U, a_{i2}^U, a_{i3}^U, a_{i4}^U, a_{i1}^L, a_{i2}^L, a_{i3}^L, a_{i4}^L$: The reference points of $\tilde{\tilde{A}}_i$,
- $H_j\left(\tilde{A}_i^U\right)$: The membership value of the $a_{i(j+1)}^U$ in the upper trapezoidal membership function $\tilde{A}_i^U$, $1 \leq j \leq 2$, and
- $H_j\left(\tilde{A}_i^L\right)$: The membership value of the $a_{i(j+1)}^L$ in the lower trapezoidal membership function $\tilde{A}_i^L$, $1 \leq j \leq 2$.

$H_1\left(\tilde{A}_i^U\right) \in [0, 1]$, $H_2\left(\tilde{A}_i^U\right) \in [0, 1]$, $H_1\left(\tilde{A}_i^L\right) \in [0, 1]$, $H_2\left(\tilde{A}_i^L\right) \in [0, 1]$, and $1 \leq i \leq n$.

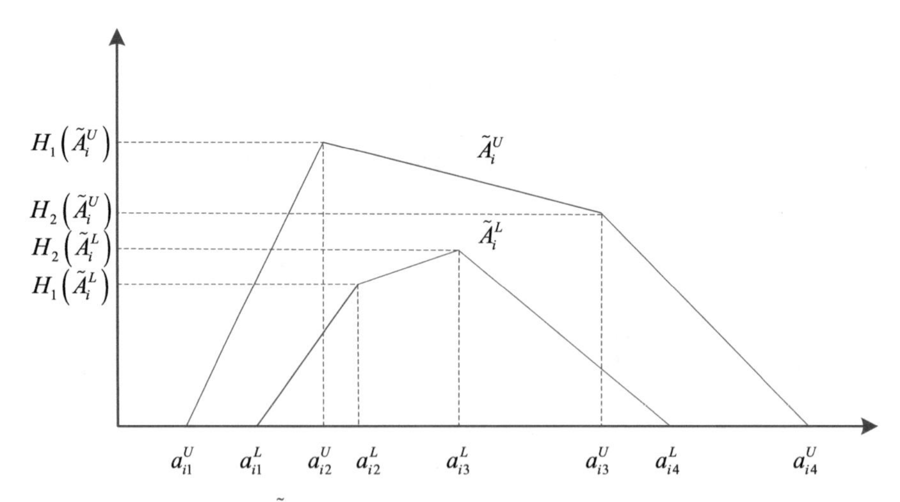


**Fig. 3.1** Graphical view of $\tilde{\tilde{A}}$ with its membership functions

### 3.1.2 *TOPSIS Method*

The traditional TOPSIS method was initially proposed by [15] as an MCDM method. According to the method, each evaluation factor has a monotonous increasing or decreasing trend. In order to create an ideal solution set, the largest of the weighted evaluation factors in the matrix is selected [16]. It uses a criteria weight matrix as well as a decision matrix which consists of alternatives, criteria, their respected performance measures.

The step-by-step procedure of the method is summarized below:

1. Define the problem and construct a decision matrix including alternatives and criteria,
2. Normalization of the decision matrix, which is provided in Step 1,
3. Make a weighted normalized decision matrix,
4. Find the ideal and the anti-ideal solution,
5. Calculate distances from the ideal and the anti-ideal solutions,
6. Calculate the relative distance of an alternative to the ideal solution, and
7. Rank the alternatives considering descending order. The highest final score shows the best alternative.

## 3.2 Proposed Fine–Kinney-Based Approach Using Interval Type-2 F-TOPSIS

**Step 1**: Let $A = \{A_1, A_2, \ldots, A_n\}$ show the alternative set, $C = \{c_1, c_2, \ldots, c_m\}$ the criteria set, and $D = \{D_1, D_2, \ldots, D_K\}$ the decision-maker set. Each decision-maker participates in the survey, which is designed for the current study. In this survey, each decision-maker evaluates each alternative (we mean "hazard" in the case study of this chapter) by giving a performance value according to each criterion (we mean "three risk parameters of Fine–Kinney in the case study of this chapter). The aggregation of each decision-maker's performance values is performed as follows:

$$E_c = (\tilde{\tilde{c}}_{ij}^{p})_{m\times n} = \begin{array}{c} \\ c_1 \\ c_2 \\ \vdots \\ c_m \end{array} \begin{array}{c} \begin{array}{cccc} A_1 & A_2 & \cdots & A_n \end{array} \\ \begin{bmatrix} \tilde{\tilde{c}}_{11}^{p} & \tilde{\tilde{c}}_{12}^{p} & \cdots & \tilde{\tilde{c}}_{1n}^{p} \\ \tilde{\tilde{c}}_{21}^{p} & \tilde{\tilde{c}}_{22}^{p} & \cdots & \tilde{\tilde{c}}_{2n}^{p} \\ \vdots & \vdots & \vdots & \vdots \\ \tilde{\tilde{c}}_{m1}^{p} & \tilde{\tilde{c}}_{m2}^{p} & \cdots & \tilde{\tilde{c}}_{mn}^{p} \end{bmatrix} \end{array} \tag{3.1}$$

where $\tilde{\tilde{c}}_{ij} = \left(\frac{\tilde{\tilde{c}}_{ij}^{1} \oplus \tilde{\tilde{c}}_{ij}^{2} \oplus \cdots \oplus \tilde{\tilde{c}}_{ij}^{k}}{k}\right)$, $1 \le i \le m$, $1 \le j \le n$, $1 \le c \le k$, and $k$ shows the number of experts.

$$\tilde{\tilde{c}}_{ij} = \left(\left(a_{i1}^{U}, a_{i2}^{U}, a_{i3}^{U}, a_{i4}^{U}; H_1\left(\tilde{A}_i^{U}\right), H_2\left(\tilde{A}_i^{U}\right)\right), \left(a_{i1}^{L}, a_{i2}^{L}, a_{i3}^{L}, a_{i4}^{L}; H_1\left(\tilde{A}_i^{L}\right), H_2\left(\tilde{A}_i^{L}\right)\right)\right)$$

**Step 2**: Achieve a weighting matrix $W_s$ for the risk parameters of Fine–Kinney by one of the common MCDM methods such as AHP, BWM, and DEMATEL or simply assigning subjectively.

$$W_s = \left(w_i^s\right)_{1\times m} = \begin{matrix} c_1 & c_2 & \cdots & c_m \\ \left[w_1^s,\right. & w_2^s, & \cdots, & \left.w_m^s\right] \end{matrix} \tag{3.2}$$

**Step 3**: Multiply the criterion weights matrix and the decision matrix to obtain the weighted decision matrix. The weighted decision matrix for each criterion is defined as follows:

$$\tilde{\tilde{v}}_{ij} = \tilde{\tilde{c}}_{ij} \times w_i \tag{3.3}$$

**Step 4**: Calculate the ranking value $Rank\left(\tilde{\tilde{v}}_{ij}\right)$ of interval type-2 fuzzy sets. The ranking weighted decision matrix $E_w$ is constructed.

**Step 5**: Determine the positive-ideal solution $x^* = \left(v_1^*, v_2^*, \ldots, v_m^*\right)$ and the negative-ideal solution $x^- = \left(v_1^-, v_2^-, \ldots, v_m^-\right)$, where

$$v_i^* = \begin{cases} \max\limits_{i\le j\le n}\left\{Rank\left(\tilde{\tilde{v}}_{ij}\right)\right\}, & \text{if } f_i \in B \\ \min\limits_{i\le j\le n}\left\{Rank\left(\tilde{\tilde{v}}_{ij}\right)\right\}, & \text{if } f_i \in C \end{cases} \tag{3.4}$$

$$v_i^- = \begin{cases} \min\limits_{i\le j\le n}\left\{Rank\left(\tilde{\tilde{v}}_{ij}\right)\right\}, & \text{if } f_i \in B \\ \max\limits_{i\le j\le n}\left\{Rank\left(\tilde{\tilde{v}}_{ij}\right)\right\}, & \text{if } f_i \in C \end{cases} \tag{3.5}$$

Then, positive-$d^*(x_j)$ and negative-$d^-(x_j)$ ideal solutions are determined and given as follows:

$$d^*(x_j) = \sqrt{\sum_{i\in I}\left(Rank(\tilde{\tilde{v}}_{ij}) - v_i^*\right)^2} \tag{3.6}$$

$$d^-(x_j) = \sqrt{\sum_{i\in I}\left(Rank(\tilde{\tilde{v}}_{ij}) - v_i^-\right)^2} \tag{3.7}$$

The ranking value $Rank(\tilde{\tilde{A}}_i)$ of $\tilde{\tilde{A}}_i$ is defined as follows:

$$\begin{aligned} Rank(\tilde{\tilde{v}}_i) = {} & M_1\left(\tilde{v}_i^U\right) + M_1\left(\tilde{v}_i^L\right) + M_2\left(\tilde{v}_i^U\right) + M_2\left(\tilde{v}_i^L\right) + M_3\left(\tilde{v}_i^U\right) + M_3\left(\tilde{v}_i^L\right) \\ & - \frac{1}{4}\left(S_1\left(\tilde{v}_i^U\right) + S_1\left(\tilde{v}_i^L\right) + S_2\left(\tilde{v}_i^U\right) + S_2\left(\tilde{v}_i^L\right) + S_3\left(\tilde{v}_i^U\right) + S_3\left(\tilde{v}_i^L\right) + S_4\left(\tilde{v}_i^U\right) + S_4\left(\tilde{v}_i^L\right)\right) \end{aligned}$$

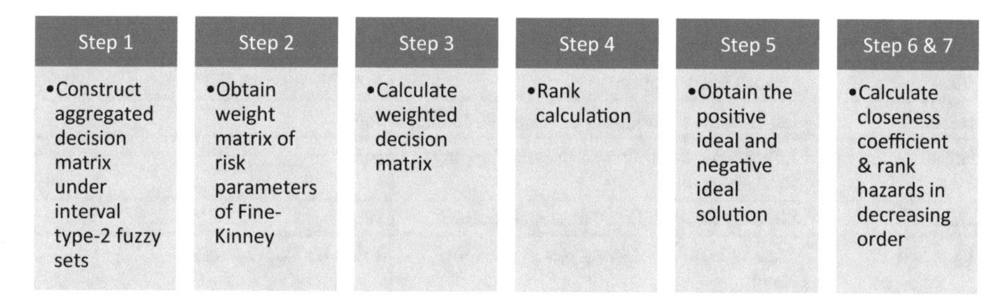


**Fig. 3.2** The step-by-step flow of the approach

$$+ H_1\left(\tilde{v}_i^U\right) + H_1\left(\tilde{v}_i^L\right) + H_2\left(\tilde{v}_i^U\right) + H_2\left(\tilde{v}_i^L\right)$$

Here the meanings and formulations of notions above can be found at [5, 6].

**Step 6**: Then, the closeness coefficient $CC(x_j)$ is computed by Eq. (3.8).

$$CC(x_j) = \frac{d^-(x_j)}{d^*(x_j) + d^-(x_j)} \tag{3.8}$$

**Step 7**: Alternatives are finally ranked in decreasing order. The larger the value of $CC(x_j)$, the best the alternative.

The main steps of the proposed approach are given in Fig. 3.2.

## 3.3 Case Study

To demonstrate the applicability of the proposed approach, a real case study is carried out for the occupational risk assessment of chrome plating unit in a gun factory. As the main method of the approach, the interval type-2 fuzzy TOPSIS method was used to rank hazards and their associated risks. In the following, step-by-step application of the proposed approach to the problem is provided.

### *3.3.1 Application Results*

In this application, three parameters of Fine–Kinney such as probability ($P$), exposure ($E$), and consequence ($C$) are used in assessing hazards. Also, the risk list regarding surface processing activities in the chrome plating unit of the observed gun factory is provided in Table 3.1. The weights of these parameters are assumed as $w_P = 0.289$, $w_E = 0.293$, $w_C = 0.418$. These values are obtained by Gul et al. [17]. During the prioritization procedure by interval type-2 fuzzy TOPSIS, the linguistic

**Table 3.1** The hazard identification list in chrome plating unit

| ID | Hazard identification | Possible effect (risk) |
|---|---|---|
| Hazard-1 | Layout of the working area (tripping, falling) | Injury |
| Hazard-2 | Insufficient working area (hitting, tripping, squeezing) | Injury |
| Hazard-3 | Slippery floor (falling, slipping, tripping) | Injury |
| Hazard-4 | Manual handling, lifting, placing, loading, pushing | Joint, lumbar disorders |
| Hazard-5 | Insufficient ventilation | Discomfort, stress, suffocation, poisoning, death |
| Hazard-6 | Continuous standing and settlement | Joint disorders, varicose veins, stress |
| Hazard-7 | Working with hand tools: (hammer, screwdriver, scissors, grinder, etc.) | Dents, cuts, injuries |
| Hazard-8 | Repetitive movements | Injury, joint disorders, stress |
| Hazard-9 | Rotating-moving parts of machines and its components (hit, squeeze, crush) | Injury, death |
| Hazard-10 | Electricity | Combustion, injury, death |
| Hazard-11 | Falling/flying objects | Injury |
| Hazard-12 | Fire | Burns, suffocation, death |
| Hazard-13 | Unsuitable climatic conditions (wind, gusts, hail, cold, heat, ice, storm, etc.) | Occupational disease, injury |
| Hazard-14 | Hot/cold spaces and surfaces, hot water | Combustion, adhesion, diseases, heat stress, death |
| Hazard-15 | Noise | Hearing loss, stress/noise pollution |
| Hazard-16 | Dust | Respiratory tract, lung diseases |
| Hazard-17 | Dump | Microbial diseases, rodents and pests, and so on. |
| Hazard-18 | Working with lifting devices (crush, squeeze, hit) | Injury, death |
| Hazard-19 | Brightness, more lighting, inadequate lighting | Eye disorders, stress |
| Hazard-20 | Energy cutting, energy supplying (electric, pneumatic, hydraulic) | Burning, shock, death |
| Hazard-21 | Exposure to chemical liquid + gas (alcohol, welding, soldering, gas, etc.) | Cancer, burns, eye ailments, irritation |
| Hazard-22 | Emergency cases (fires, earthquakes, floods, etc.) | Injury, death |
| Hazard-23 | Professional qualifications/Experience | Injury, death |

terms given in Table 3.2 are used. This scale is obtained from Celik et al. [5]. While using the terms from the table, the evaluations given in Table 3.3 are obtained to construct the fuzzy aggregated decision matrix.

**Table 3.2** The scale used in assessing hazards*

| Linguistic term | Interval type-2 fuzzy numbers |
|---|---|
| Poor (P) | ((0, 1, 1, 3; 1, 1), (0.5, 1, 1, 2; 0.9, 0.9)) |
| Medium Poor (MP) | ((1, 3, 3, 5; 1, 1), (2, 3, 3, 4; 0.9, 0.9)) |
| Medium (M) | ((3, 5, 5, 7; 1, 1), (4, 5, 5, 6; 0.9, 0.9)) |
| Medium Good (MG) | ((5, 7, 7, 9; 1, 1), (6, 7, 7, 8; 0.9, 0.9)) |
| Good (G) | ((7, 9, 9, 10; 1, 1), (8, 9, 9, 9.5; 0.9, 0.9)) |
| Very Good (VG) | ((9, 10, 10, 10; 1, 1), (9.5, 10, 10, 10; 0.9, 0.9)) |

**Table 3.3** The linguistic initial decision matrix from the OHS experts' consensus

| Hazard ID | Probability | Exposure | Consequence |
|---|---|---|---|
| Hazard-1 | MG | G | M |
| Hazard-2 | MG | G | M |
| Hazard-3 | G | G | M |
| Hazard-4 | MG | G | M |
| Hazard-5 | G | VG | MG |
| Hazard-6 | MG | VG | M |
| Hazard-7 | MG | MG | M |
| Hazard-8 | MG | VG | M |
| Hazard-9 | MG | G | MG |
| Hazard-10 | M | MP | G |
| Hazard-11 | M | G | M |
| Hazard-12 | M | VG | MG |
| Hazard-13 | M | MP | MP |
| Hazard-14 | MG | G | M |
| Hazard-15 | MG | VG | MG |
| Hazard-16 | M | MP | M |
| Hazard-17 | M | MP | MP |
| Hazard-18 | MG | MG | MG |
| Hazard-19 | MG | MP | MG |
| Hazard-20 | M | G | G |
| Hazard-21 | MG | G | MG |
| Hazard-22 | M | MG | MG |
| Hazard-23 | MG | G | MG |

By following the steps as indicated in Sect. 3.2, we obtain the final values of the interval type-2 fuzzy TOPSIS model. Using the model, $d^*(x_j)$, $d^-(x_j)$, and $CC(x_j)$ values are calculated and hazards are ranked according to the decreasing order of $CC(x_j)$ values (Table 3.4). The results of the study demonstrate that the most important three risks are insufficient ventilation (Hazard-5), noise (Hazard-15) and energy cutting, energy supplying (electric, pneumatic, hydraulic) (Hazard-20).

**Table 3.4** Final risk scores and rankings by the proposed approach

| Hazard | $d^*(x_j)$ | $d^-(x_j)$ | $CC(x_j)$ | Rank | Hazard | $d^*(x_j)$ | $d^-(x_j)$ | $CC(x_j)$ | Rank |
|---|---|---|---|---|---|---|---|---|---|
| Hazard-1 | 1.119 | 1.457 | 0.566 | 10 | Hazard-13 | 4.200 | 0.000 | 0.000 | 16 |
| Hazard-2 | 1.119 | 1.457 | 0.566 | 10 | Hazard-14 | 1.119 | 1.457 | 0.566 | 10 |
| Hazard-3 | 1.007 | 1.801 | 0.641 | 7 | Hazard-15 | 0.345 | 2.654 | 0.885 | 2 |
| Hazard-4 | 1.119 | 1.457 | 0.566 | 10 | Hazard-16 | 2.961 | 0.252 | 0.078 | 15 |
| Hazard-5 | 0.233 | 2.997 | 0.928 | 1 | Hazard-17 | 4.200 | 0.000 | 0.000 | 16 |
| Hazard-6 | 1.081 | 1.899 | 0.637 | 8 | Hazard-18 | 0.629 | 1.621 | 0.721 | 6 |
| Hazard-7 | 1.365 | 0.866 | 0.388 | 13 | Hazard-19 | 1.872 | 1.127 | 0.376 | 14 |
| Hazard-8 | 1.081 | 1.899 | 0.637 | 8 | Hazard-20 | 0.501 | 3.295 | 0.868 | 3 |
| Hazard-9 | 0.383 | 2.212 | 0.853 | 4 | Hazard-21 | 0.383 | 2.212 | 0.853 | 4 |
| Hazard-10 | 1.991 | 2.209 | 0.526 | 11 | Hazard-22 | 0.981 | 1.501 | 0.605 | 9 |
| Hazard-11 | 1.471 | 1.337 | 0.476 | 12 | Hazard-23 | 0.383 | 2.212 | 0.853 | 4 |
| Hazard-12 | 0.697 | 2.534 | 0.784 | 5 | | | | | |

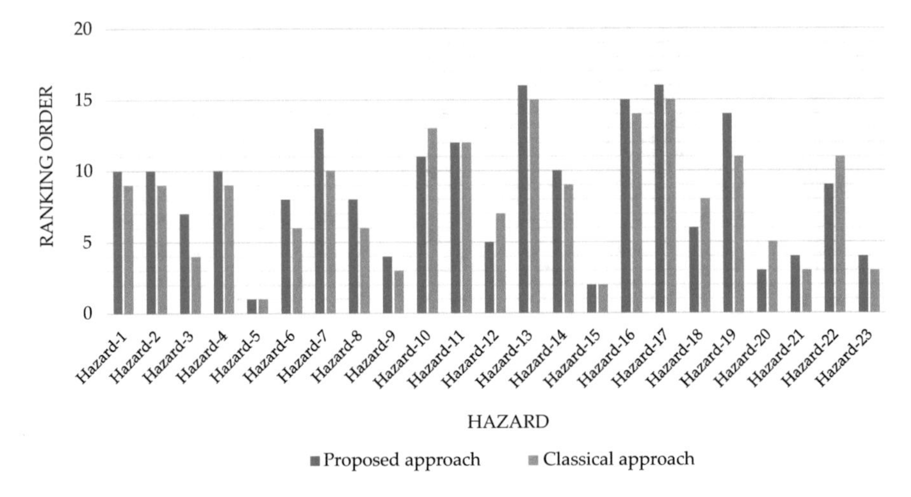


**Fig. 3.3** Comparison of rankings by the proposed and the classic approach

### 3.3.2 Validation Study on the Results

In this subsection, some validation tests of the obtained ranking results are provided. As a first validation study, we made a comparative study between the results of the existed approach (interval type-2 fuzzy TOPSIS under Fine–Kinney's method) and the classical Fine–Kinney method. We then observe the variations in hazard rankings. The results are shown in Fig. 3.3.

**Table 3.5** The weight vectors designed for the sensitivity analysis

| Weight vector | Parameter | Weight value |
|---|---|---|
| Weight vector-1 ($W_1$) | Probability | 0.289 |
| | Exposure | 0.293 |
| | Consequence | 0.418 |
| Weight vector-2 ($W_2$) | Probability | 0.250 |
| | Exposure | 0.200 |
| | Consequence | 0.550 |
| Weight vector-3 ($W_3$) | Probability | 0.333 |
| | Exposure | 0.333 |
| | Consequence | 0.333 |
| Weight vector-4 ($W_4$) | Probability | 0.400 |
| | Exposure | 0.400 |
| | Consequence | 0.200 |

It is observed from Fig. 3.3 that by both approaches, Hazard-5 is ranked as the most critical hazard, followed by Hazard-15. It is also seen that the least important three hazards (Hazard-13, Hazard 16, and Hazard 17) are the same according to both approaches. When we compare the results obtained by both approaches, we observe that there are very small rank variations between them. The Spearman rank correlation between the two approaches is obtained as 0.943. That means there exists a high correlation with the ranking orders of two approaches. So, it can be claimed that this proposed approach is applicable for occupational risk assessment in the Fine–Kinney domain.

As a second validation study, we made a sensitivity analysis and analyze the variation in hazard ranking with respect to the changes in parameters' weights. Therefore, we apply different weight vectors of the Fine–Kinney parameters. A total of four combinations are created. The weight vectors of Fine–Kinney parameters designed for the sensitivity analysis are given in Table 3.5. The ranking orders with respect to the sensitivity analysis are shown in Table 3.6.

It can be observed from Table 3.6 that when the weight vector varies, there are variations in the ranking orders of hazards. Hence, our method is sensitive to Fine–Kinney risk parameters' weights. Hazard-5 is mostly ranked as the most critical hazard. It has the second ranking according to the second weight vector ($W_2$). There is no change in the ranking of Hazard-13, Hazard-16, and Hazard-17 for all combinations. When compared to the results with the ones similar to this study from the literature, we can say that the ranking result obtained by our proposed approach is credible and applicable.

**Table 3.6** Ranking order changes in times of parameters' weights changes

| Hazard | Ranking order of the hazard | | | |
|---|---|---|---|---|
| | $W_1$ | $W_2$ | $W_3$ | $W_4$ |
| Hazard-1 | 10 | 12 | 9 | 6 |
| Hazard-2 | 10 | 12 | 9 | 6 |
| Hazard-3 | 7 | 10 | 6 | 3 |
| Hazard-4 | 10 | 12 | 9 | 6 |
| Hazard-5 | 1 | 2 | 1 | 1 |
| Hazard-6 | 8 | 11 | 7 | 5 |
| Hazard-7 | 13 | 14 | 12 | 10 |
| Hazard-8 | 8 | 11 | 7 | 5 |
| Hazard-9 | 4 | 4 | 3 | 4 |
| Hazard-10 | 11 | 5 | 13 | 13 |
| Hazard-11 | 12 | 13 | 10 | 9 |
| Hazard-12 | 5 | 7 | 5 | 7 |
| Hazard-13 | 16 | 16 | 16 | 15 |
| Hazard-14 | 10 | 12 | 9 | 6 |
| Hazard-15 | 2 | 3 | 2 | 2 |
| Hazard-16 | 15 | 15 | 15 | 14 |
| Hazard-17 | 16 | 16 | 16 | 15 |
| Hazard-18 | 6 | 6 | 8 | 9 |
| Hazard-19 | 14 | 9 | 14 | 12 |
| Hazard-20 | 3 | 1 | 4 | 8 |
| Hazard-21 | 4 | 4 | 3 | 4 |
| Hazard-22 | 9 | 8 | 11 | 11 |
| Hazard-23 | 4 | 4 | 3 | 4 |

## 3.4 Python Implementation of the Proposed Approach

```
# Chapter 3
# import required libraries
import numpy as np

# set initial variables
n_criteria = 3
n_hazard = 23
n_fuzzy_scores = 6 # number of linguistics
n_fuzzy_values = 12
fn_score = 10 # fuzzy normalization score
criteria_weight = [0.4, 0.4, 0.2] # P, E, C
fuzzy_scores = [
  ["P", 0, 1, 1, 3, 1, 1, 0.5, 1, 1, 2, 0.9, 0.9], # Poor (P)
  ["MP", 1, 3, 3, 5, 1, 1, 2, 3, 3, 4, 0.9, 0.9], # Medium Poor (MP)
  ["M", 3, 5, 5, 7, 1, 1, 4, 5, 5, 6, 0.9, 0.9], # Medium (M)
  ["MG", 5, 7, 7, 9, 1, 1, 6, 7, 7, 8, 0.9, 0.9], # Medium Good (MG)
  ["G", 7, 9, 9, 10, 1, 1, 8, 9, 9, 9.5, 0.9, 0.9], # Good (G)
  ["VG", 9, 10, 10, 10, 1, 1, 9.5, 10, 10, 10, 0.9, 0.9] # Very Good (VG)
  ]
  # ehe: 1st expert's hazard evaluation
  ehe = [
    # Hazard ID, Probability, Exposure, Consequence
    ["Hazard-1", "MG", "G", "M"],
    ["Hazard-2", "MG", "G", "M"],
    ["Hazard-3", "G", "G", "M"],
    ["Hazard-4", "MG", "G", "M"],
    ["Hazard-5", "G", "VG", "MG"],
    ["Hazard-6", "MG", "VG", "M"],
    ["Hazard-7", "MG", "MG", "M"],
    ["Hazard-8", "MG", "VG", "M"],
    ["Hazard-9", "MG", "G", "MG"],
    ["Hazard-10", "M", "MP", "G"],
    ["Hazard-11", "M", "G", "M"],
    ["Hazard-12", "M", "VG", "MG"],
    ["Hazard-13", "M", "MP", "MP"],
    ["Hazard-14", "MG", "G", "M"],
    ["Hazard-15", "MG", "VG", "MG"],
    ["Hazard-16", "M", "MP", "M"],
    ["Hazard-17", "M", "MP", "MP"],
    ["Hazard-18", "MG", "MG", "MG"],
    ["Hazard-19", "MG", "MP", "MG"],
    ["Hazard-20", "M", "G", "G"],
    ["Hazard-21", "MG", "G", "MG"],
    ["Hazard-22", "M", "MG", "MG"],
    ["Hazard-23", "MG", "G", "MG"]
  ]
```

```
def rank(vector, da): # da -1:descending, 1:ascending
  order = np.zeros([len(vector), 1])
  unique_val = da * np.sort(da * np.unique(vector))
  for ix in range(0, len(unique_val)):
    order[np.argwhere(vector == unique_val[ix])] = ix +1
  return order

def print_result(order, vector):
  print('Hazard Id, Rank, Value')
  for ix in range(0, len(order)):
    print(ehe[ix][0], ', ', int(order[ix]), ', ', vector[ix])

# nfs: normalize fuzzy scores
for i in range(0, n_fuzzy_scores):
  fuzzy_scores[i][1:5] = np.asarray(fuzzy_scores[i][1:5]) / fn_score
  fuzzy_scores[i][7:11] = np.asarray(fuzzy_scores[i][7:11]) / fn_score

# calculate weighted decision matrix (wdm)
wdm = np.zeros([n_hazard, n_criteria, n_fuzzy_values])
for hz_ix in range(0, n_hazard):
  for cr_ix in range(0, n_criteria):
    for fs_ix in range(0, n_fuzzy_scores):
      if ehe[hz_ix][cr_ix + 1] == fuzzy_scores[fs_ix][0]:
        for fv_ix in range(0, n_fuzzy_values):
          if fv_ix ==4 or fv_ix ==5 or fv_ix ==10 or fv_ix ==11:
            temp_val = 1
          else:
            temp_val = criteria_weight[cr_ix]
          wdm[hz_ix][cr_ix][fv_ix] = fuzzy_scores[fs_ix][fv_ix +1] * temp_val
# generate a temporary matrix (tm) for calculations
tm = np.zeros([n_hazard, n_criteria])
for ix_hz in range(0, n_hazard):
  for ix_cr in range(0, n_criteria):
    for ix in [0, 6]:
      tm[ix_hz][ix_cr] += \
        np.mean(wdm[ix_hz][ix_cr][ix + 0:ix + 2]) \
        + np.mean(wdm[ix_hz][ix_cr][ix +1:ix + 3]) \
        + np.mean(wdm[ix_hz][ix_cr][ix +2:ix + 4]) \
        - np.mean(
          [np.std(wdm[ix_hz][ix_cr][ix + 0:ix + 2]),
           np.std(wdm[ix_hz][ix_cr][ix + 1:ix + 3]),
           np.std(wdm[ix_hz][ix_cr][ix + 2:ix + 4]),
           np.std(wdm[ix_hz][ix_cr][ix + 0:ix + 4])]
        ) + sum(wdm[ix_hz][ix_cr][ix + 4:ix + 6])
```

```
max_val, min_val = np.max(tm, axis=0), np.min(tm, axis=0)
max_rank = np.empty([n_hazard, n_criteria])
min_rank = np.empty([n_hazard, n_criteria])
max_rank = np.square(tm - max_val)
min_rank = np.square(tm - min_val)
D_plus = np.sum(max_rank, axis=1)
D_minus = np.sum(min_rank, axis=1)
CC = D_minus / (D_plus + D_minus)
hazard_rank = rank(CC, 1)
print_result(hazard_rank, CC)
'''
Output:
Hazard Id, Rank, Value
Hazard-1 , 10 , 0.565669345018739
Hazard-2 , 10 , 0.565669345018739
Hazard-3 , 7 , 0.6412477330842311
Hazard-4 , 10 , 0.565669345018739
Hazard-5 , 1 , 0.927741429902587
Hazard-6 , 8 , 0.6372032432920185
Hazard-7 , 13 , 0.388243395883186
Hazard-8 , 8 , 0.6372032432920185
Hazard-9 , 4 , 0.8525216733027218
Hazard-10 , 11 , 0.5259885561288329
Hazard-11 , 12 , 0.4761650095188426
Hazard-12 , 5 , 0.7842590254541644
Hazard-13 , 16 , 0.0
Hazard-14 , 10 , 0.565669345018739
Hazard-15 , 2 , 0.8849496626772264
Hazard-16 , 15 , 0.07832302866307285
Hazard-17 , 16 , 0.0
Hazard-18 , 6 , 0.7205268214954427
Hazard-19 , 14 , 0.37566647762494376
Hazard-20 , 3 , 0.8679654620144249
Hazard-21 , 4 , 0.8525216733027218
Hazard-22 , 9 , 0.604791136322043
Hazard-23 , 4 , 0.8525216733027218
'''
```

## References

1. Hwang, C.L, Yoon, K. (1981). *Multiple attribute decision making: Methods and applications, a state of the art survey*. Springer, New York.
2. Mendel, J. M., John, R. I., & Liu, F. (2006). Interval type-2 fuzzy logic systems made simple. *IEEE Transactions on Fuzzy Systems, 14*(6), 808–821.
3. Lee, L. W., & Chen, S. M. (2008). Fuzzy multiple attributes group decision-making based on the extension of TOPSIS method and interval type-2 fuzzy sets. In *International Conference on Machine Learning and Cybernetics, 2008* (Vol. 6, pp. 3260–3265). IEEE.

4. Chen, S. M., & Lee, L. W. (2010). Fuzzy multiple attributes group decision-making based on the interval type-2 TOPSIS method. *Expert Systems with Applications, 37*(4), 2790–2798.
5. Celik, E., Aydin, N., & Gumus, A. T. (2014). A multiattribute customer satisfaction evaluation approach for rail transit network: A real case study for Istanbul, Turkey. *Transport Policy, 36,* 283–293.
6. Celik, E., Bilisik, O. N., Erdogan, M., Gumus, A. T., & Baracli, H. (2013). An integrated novel interval type-2 fuzzy MCDM method to improve customer satisfaction in public transportation for Istanbul. *Transportation Research Part E: Logistics and Transportation Review, 58,* 28–51.
7. Celik, E., Gul, M., Aydin, N., Gumus, A. T., & Guneri, A. F. (2015). A comprehensive review of multi criteria decision making approaches based on interval type-2 fuzzy sets. *Knowledge-Based Systems, 85,* 329–341.
8. Celik, E., Gumus, A. T., & Erdogan, M. (2016). A new extension of the ELECTRE method based upon interval type-2 fuzzy sets for green logistic service providers evaluation. *Journal of Testing and Evaluation, 44*(5), 1813–1827.
9. Soner, O., Celik, E., & Akyuz, E. (2017). Application of AHP and VIKOR methods under interval type 2 fuzzy environment in maritime transportation. *Ocean Engineering, 129,* 107–116.
10. Demirel, H., Akyuz, E., Celik, E., & Alarcin, F. (2019). An interval type-2 fuzzy QUALIFLEX approach to measure performance effectiveness of ballast water treatment (BWT) system onboard ship. *Ships and Offshore Structures, 14*(7), 675–683.
11. Celik, E., & Gumus, A. T. (2016). An outranking approach based on interval type-2 fuzzy sets to evaluate preparedness and response ability of non-governmental humanitarian relief organizations. *Computers & Industrial Engineering, 101,* 21–34.
12. Celik, E., & Gumus, A. T. (2018). An assessment approach for non-governmental organizations in humanitarian relief logistics and an application in Turkey. *Technological and Economic Development of Economy, 24*(1), 1–26.
13. Kahraman, C., Öztayşi, B., Sarı, İ. U., & Turanoğlu, E. (2014). Fuzzy analytic hierarchy process with interval type-2 fuzzy sets. *Knowledge-Based Systems, 59,* 48–57.
14. Celik, E. (2017). A cause and effect relationship model for location of temporary shelters in disaster operations management. *International Journal of Disaster Risk Reduction, 22,* 257–268.
15. Yoon, K. P., & Hwang, C. L. (1995). *Multiple attribute decision making: An introduction* (Vol. 104). Thousand Oaks, CA, USA, Sage Publications.
16. Oz, N. E., Mete, S., Serin, F., & Gul, M. (2019). Risk assessment for clearing and grading process of a natural gas pipeline project: An extended TOPSIS model with Pythagorean fuzzy sets for prioritizing hazards. *Human and Ecological Risk Assessment: An International Journal, 25*(6), 1615–1632.
17. Gul, M., Guven, B., & Guneri, A. F. (2018). A new Fine-Kinney-based risk assessment framework using FAHP-FVIKOR incorporation. *Journal of Loss Prevention in the Process Industries, 53,* 3–16.

# Chapter 4
# Fine–Kinney-Based Occupational Risk Assessment Using Interval-Valued Pythagorean Fuzzy VIKOR

**Abstract** This chapter aims at the adaptation of the Fine–Kinney occupational risk assessment concept into the VIKOR multi-attribute decision-making method with an interval-valued Pythagorean fuzzy set. The classical fuzzy set theory has been improved by proposing a number of extended versions. One of them is the Pythagorean fuzzy set. It has been firstly developed by Yager (IEEE Transactions on Fuzzy Systems 22:958–965, [1]). In this chapter, we use this type of fuzzy set with VIKOR since it reflects uncertainty in occupational risk assessment and decision-making better than other fuzzy extensions. To demonstrate the proposed approach applicability, a case study regarding the activities of the surface treatment area in a chrome plating unit of a gun factory is performed. Some additional analysis to test the solidity and validity of the approach is executed. Finally, the Python codes in the implementation of the proposed approach are given for scholars and practitioners for usage in further studies.

## 4.1 Pythagorean Fuzzy Sets and VIKOR

In this part, before explaining the proposed approach, some preliminaries are presented regarding Pythagorean fuzzy sets and more specifically interval-valued Pythagorean fuzzy sets. Then, the algorithm interval-valued Pythagorean fuzzy VIKOR (IVPFVIKOR) is provided in detail.

### *4.1.1 General View on Pythagorean Fuzzy Sets*

Pythagorean fuzzy sets have been first of all suggested by Yager [1] and have been applied by many scholars to various areas to handle uncertainty such as intuitionistic fuzzy sets. These two of the sets include a membership, nonmembership, and hesitancy degree. On the other hand, the criterion of membership and nonmembership degrees that are larger than 1 cannot be satisfied by intuitionistic fuzzy sets.



M. Gul et al., *Fine–Kinney-Based Fuzzy Multi-criteria Occupational Risk Assessment*, Studies in Fuzziness and Soft Computing 398,
https://doi.org/10.1007/978-3-030-52148-6_4

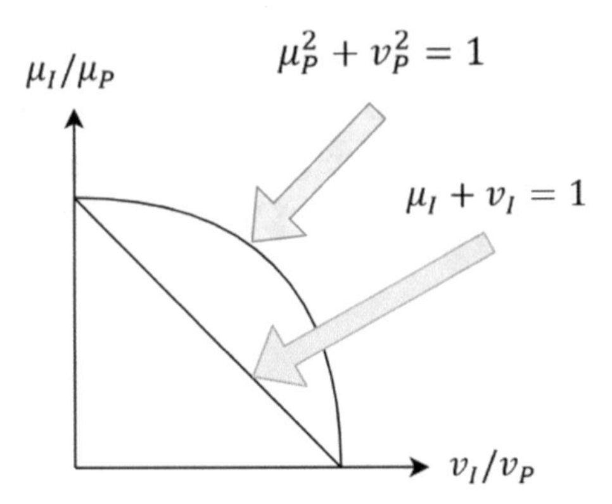


**Fig. 4.1** Pythagorean and intuitionistic fuzzy number comparison according to spaces Adapted from Ref. [17], with kind permission from Springer Science + Business Media

To overcome this drawback, Pythagorean fuzzy sets are developed [1]. This type of fuzzy sets is considered as flexible and more powerful to resolve issues regarding uncertainty [2–16].

The sum of membership and nonmembership degrees can exceed 1 while the sum of squares cannot surpass 1 in Pythagorean fuzzy sets [2–6, 15–17]. This situation is given in Definition (1) and Fig. 4.1.

**Definition 1** Let $X$ be a set in a discourse universe. A Pythagorean fuzzy set $P$ has the form in the following [16]:

$$P = \{< x,\ P(\mu_P(x), v_P(x)) > | x \in X\} \tag{4.1}$$

where $\mu_P(x) : X \mapsto [0, 1]$ describes the degree of membership and $v_P(x) : X \mapsto [0, 1]$ describes the degree of nonmembership of the element $x \in X$ to $P$, respectively, and, for every $x \in X$, it takes

$$0 \le \mu_P(x)^2 + v_P(x)^2 \le 1 \tag{4.2}$$

For any Pythagorean fuzzy set $P$ and $x \in X$, $\pi_P(x) = \sqrt{1 - \mu_P^2(x) - v_P^2(x)}$ is named as the degree of indeterminacy of $x$ to $P$.

**Definition 2** Let $P_1 = P(\mu_{P_1}, v_{P_1})$ and $P_2 = P(\mu_{P_2}, v_{P_2})$ be two Pythagorean fuzzy numbers, and $\lambda > 0$, then the operations on these two Pythagorean fuzzy numbers are determined as shown below [15, 16, 18, 19]:

$$P_1 \oplus P_2 = P\left(\sqrt{\mu_{P_1}^2 + \mu_{P_2}^2 - \mu_{P_1}^2\mu_{P_2}^2}, v_{P_1}v_{P_2}\right) \tag{4.3}$$

$$P_1 \otimes P_2 = P\left(\mu_{P_1}\mu_{P_2}, \sqrt{v_{P_1}^2 + v_{P_2}^2 - v_{P_1}^2 v_{P_2}^2}\right) \tag{4.4}$$

$$\lambda P_1 = P\left(\sqrt{1-(1-\mu_{P_1}^2)^\lambda}, (v_{P_1})^\lambda\right), \ \lambda > 0 \tag{4.5}$$

$$P_1^\lambda = P\left((\mu_{P_1})^\lambda, \sqrt{1-(1-v_{P_1}^2)^\lambda}\right)\lambda > 0 \tag{4.6}$$

**Definition 3** Let $P_1 = P(\mu_{P_1}, v_{P_1})$ and $P_2 = P(\mu_{P_2}, v_{P_2})$ be two Pythagorean fuzzy numbers, a nature quasi-ordering on the Pythagorean fuzzy numbers is determined as shown below [16]:

$$P_1 \geq P_2 \text{ if and only if } \mu_{P_1} \geq \mu_{P_2} \text{ and } v_{P_1} \leq v_{P_2}$$

A score function is developed by Zhang and Xu [16] to compare two Pythagorean fuzzy numbers of magnitude given as follows:

$$s(P_1) = (\mu_{P_1})^2 - (v_{P_1})^2 \tag{4.7}$$

**Definition 4** For the Pythagorean fuzzy numbers which are given above according to the proposed score functions, to compare two Pythagorean fuzzy numbers, the following laws are defined [16]:

(1) If $s(P_1) < s(P_2)$, then $P_1 \prec P_2$
(2) If $s(P_1) > s(P_2)$, then $P_1 \succ P_2$
(3) If $s(P_1) = s(P_2)$, then $P_1 \sim P_2$

### *4.1.2 VIKOR Method*

The term "VIKOR" is originally in Serbian (VlseKriterijumska Optimizacija I Kompromisno Resenje) and means multi-criteria optimization and a compromise solution. It has been initially proposed by Opricovic [20]. It needs a criteria weight matrix and a decision matrix that cover alternatives, criteria, and their respective performance measures (values of alternatives with respect to the criteria). The procedural steps of VIKOR are specified as follows [21]:

1. Identify the problem and build a payoff (decision) matrix,
2. Define the best and the worst values of all criterion functions,
3. Compute *S* and *R* values that are specific for VIKOR,
4. Compute *Q* values according to the computed S and R values from Step 2,
5. Sort alternatives in descending order by *S*, *R*, and *Q* values, and
6. Propose a compromise solution provided two conditions (acceptable stability and acceptable advantage) are fulfilled.

## 4.2 Proposed Fine–Kinney-Based Approach Using IVPFVIKOR

In this chapter, the problem has $t$ OHS experts $E_m(m = 1 \text{ to } t)$, $f$ hazards $H_a(a = 1 \text{ to } f)$, and $s$ Fine–Kinney risk parameters $RP_z(z = 1 \text{ to } s)$. Each OHS expert $E_m$ has a weight value ($w_m > 0$ and $\sum w_m = 1$). In the lights of this initial notations and indices, the application phases of IVPFVIKOR are provided as follows:

**Step 1**: In the first phase, the Pythagorean fuzzy decision matrix with respect to the OHS experts' subjective judgments is constructed. In evaluating the hazards by the OHS experts, the seven-point Pythagorean fuzzy linguistic scale of Yazdi [22] is used. Each OHS expert's judgment is combined into a group consensus to set up the decision matrix in Pythagorean fuzzy numbers.

Let $\widetilde{r}^k_{az} = \left(\mu^k_{az}, v^k_{az}\right)$ be the Pythagorean fuzzy values provided by $E_m$ on the assessment of $H_a$ with respect to $RP_z$.

After that, the Pythagorean fuzzy hazard ratings $\left(\widetilde{r}^k_{az}\right)$ according to each of the risk parameters are computed with a Pythagorean fuzzy weighted averaging (PFWA) operator of Yazdi [22].

$$\begin{aligned}\widetilde{r}_{az} &= \text{PFWA}\left(\widetilde{r}^1_{az}, \widetilde{r}^2_{az}, \ldots, \widetilde{r}^t_{az}\right) = \oplus^t_{m=1}\lambda_m \widetilde{r}^m_{az} \\ &= \left(\sqrt{1 - \prod_{m=1}^{t}\left(1 - \left(\mu^m_{az}\right)^2\right)^{w_m}}, \prod_{m=1}^{t}\left(v^m_{az}\right)^{w_m}\right) a = 1, 2, \ldots, f, z = 1, 2, \ldots, s\end{aligned} \tag{4.8}$$

Then, the problem is formed into a matrix form as in Eq. (9):

$$\tilde{R} = \begin{bmatrix} \widetilde{r}_{11} & \cdots & \widetilde{r}_{1s} \\ \vdots & \ddots & \vdots \\ \widetilde{r}_{f1} & \cdots & \widetilde{r}_{fs} \end{bmatrix} \tag{4.9}$$

where $\widetilde{r}_{az} = (\mu_{az}, v_{az})$. is an element of the aggregated Pythagorean fuzzy decision matrix $\tilde{R}$.

**Step 2**: In the second phase, the Pythagorean fuzzy positive and negative ideal solutions, which are (PFPIS) $\tilde{p}^*_z = \left(\mu^*_z, v^*_z\right)$. and (PFNIS) $\tilde{p}^-_z = \left(\mu^-_z, v^-_z\right)$, respectively, are determined as follows:

$$\tilde{p}^*_z = \begin{cases} \max\limits_a \widetilde{r}_{az} \text{for benefit criteria} \\ \min\limits_a \widetilde{r}_{az} \text{for cost criteria} \end{cases} \quad z = 1, 2, \ldots, s$$

$$\tilde{p}^-_z = \begin{cases} \min\limits_a \widetilde{r}_{az} \text{for benefit criteria} \\ \max\limits_a \widetilde{r}_{az} \text{for cost criteria} \end{cases} \quad z = 1, 2, \ldots, s \tag{4.11}$$

**Step 3**: In the third phase, $S_a$ and $R_a$ which are both VIKOR-specific scores as formulated and calculated with the aid of generalized Pythagorean fuzzy ordered weighted standardized distance operator (GPFOWSD) of Yazdi [22] are

$$S_a = \text{GPFOWSD}\left(\tilde{p}_1^*, \tilde{p}_1^-, \tilde{r}_{a1}, \ldots, \tilde{p}_1^*, \tilde{p}_1^-, \tilde{r}_{as}\right) = \left(\sum_{m=1}^{s} w_m \bar{d}_m^{\lambda}\right)^{1/\lambda}, a = 1, 2, \ldots, f \tag{4.12}$$

$$R_a = \left(\max_m\left(w_m \bar{d}_m^{\lambda}\right)\right)^{1/\lambda}, a = 1, 2, \ldots, f \tag{4.13}$$

where $w_m$ are Fine–Kinney risk parameters' ordered weights.

**Step 4**: In this fourth phase, the index of "$Q_a$" is calculated as follows:

$$Q_a = v\frac{S_a - S^*}{S^- - S^*} + (1 - v)\frac{R_a - R^*}{R^- - R^*} \qquad a = 1, 2, \ldots, f \tag{4.14}$$

where $S^* = \min_a S_a, S^- = \max_a S_a, R^* = \min_a R_a, R^- = \max_a R_a$.

$v$ is the maximum group utility weight, whereas $(1-v)$ is the individual regret weight. Mostly, $v$ is set to 0.5.

**Step 5**: In the fifth phase, ranks of hazards are determined considering $S_a$, $R_a$ and $Q_a$ values in increasing order.

**Step 6**: The last phase concerns the conditions of the compromise solution of VIKOR. The alternative ($A^{(1)}$) which was best ordered by the measure $Q_a$ was recommended if the conditions in [20, 21] were satisfied. The main steps of the proposed approach are graphically demonstrated in Fig. 4.2.

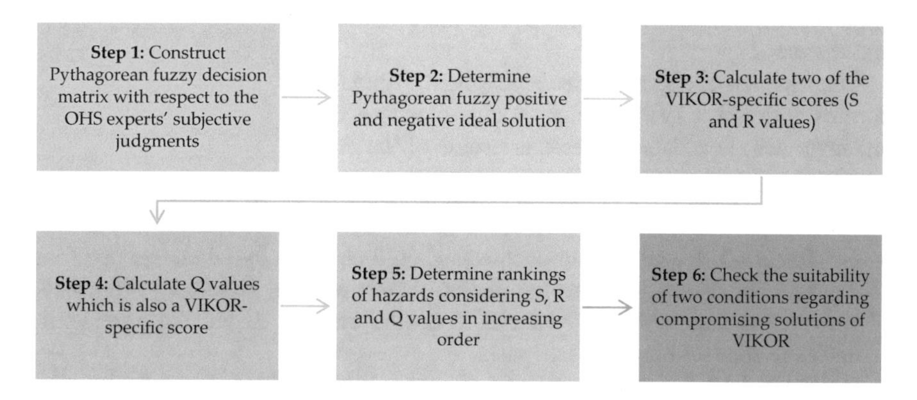


**Fig. 4.2** The main steps of the proposed approach

## 4.3 Case Study

To show the proposed approach applicability, a case study was executed in the chrome plating unit of a gun factory. The hazards and associated risks regarding surface treatment are analyzed. Three experts involved in assessing occupational hazard risks in the study. Different weights are assigned for each OHS expert in risk assessment. These OHS experts are denoted as E-1, E-2, and E-3. The weights of experts are assigned considering their work experience period in the worksite following the computation procedure of [23, 24]. The assigned weights are 0.4, 0.3, and 0.3, respectively. 23 different hazards have an impact on the safety risk of the observed gun production company. The hazard list is demonstrated in Table 4.1.

### *4.3.1 Application Results*

In this application, as in other chapters, three parameters of Fine–Kinney method are used in the risk assessment. The weights of these parameters are derived from [25] as $w_P = 0.289$, $w_E = 0.293$, $w_C = 0.418$. During the prioritization procedure by IVPFVIKOR, the linguistic terms given in [22] are used. In the chapter, the OHS experts' evaluations in Pythagorean linguistic terms for each of the 23 hazards have been received. At the end of this assessment, the linguistic evaluations of the hazards and Pythagorean fuzzy decision matrix are obtained by Eq. (8). The results are shown in Tables 4.2 and 4.3.

Then, employing Eqs. (12–14), $Q$ values are obtained as given in Table 4.4. Figure 4.3 also indicates the values of $S$, $R$, and $Q$ for each hazard. The hazards are ordered considering the $Q$ values. The smallest $Q$ value refers to the highest and most serious risk. Risks with the highest $Q$ value indicate the least important risks. Results show that the most serious risks are stemmed from *Hazard-5*, *Hazard-15*, and *Hazard-23*.

The last step of a generic risk assessment work is risk control. For this aim, for hazards with lower IVPFVIKOR-specific $Q$ values, a number of control measures are suggested. For instance, these are some of the measures that can be applied to control the system regarding *Hazard-5* (regarding insufficient ventilation), *Hazard-15* (Noise), and *Hazard-23* (professional competence and experience):

- For *Hazard-5*: Required improvements should be made by taking service from a professional company for ventilation measurements. Especially systematic suction ventilation should be installed on the benches. It will be provided to enter the zone with a half-face mask.
- For *Hazard-15*: Appropriate ear protectors should be provided and used depending on the results of the ambient measurements made to determine the noise levels of work equipment during operation. Employees should be trained about the usefulness and requirement of personal protective equipment usage.
- For *Hazard-23*: Employees should be given vocational training and be certified.

**Table 4.1** Hazards in the observed chrome plating unit of the gun factory

| Code | Hazard identification | Possible effect |
|---|---|---|
| Hazard-1 | Tripping hazard regarding the layout of the worksite | Wound and bruise, injury |
| Hazard-2 | Insufficient worksite | Injuries |
| Hazard-3 | Slippery floor | Injuries |
| Hazard-4 | Manual handling, lifting, placing, loading, and forcing | Joint, waist discomfort |
| Hazard-5 | Insufficient ventilation | Discomfort, stress, poisoning, and death |
| Hazard-6 | Constant standing, constant sitting | Joint disorders, varicose veins, and stress |
| Hazard-7 | Working with the hand tool | Cuts, injuries |
| Hazard-8 | Repeated movements | Injury, joint ailments, and stress |
| Hazard-9 | Rotating-moving parts of the machine | Injury, death |
| Hazard-10 | Electricity | Combustion, injury, and death |
| Hazard-11 | Falling and flying objects | Injuries |
| Hazard-12 | Fire | Burn, drowning, and death |
| Hazard-13 | Unsuitable climatic conditions | Disease, injury |
| Hazard-14 | Hot/cold places and surfaces | Burning, sticking, diseases, heat stress, and death |
| Hazard-15 | Noise | Hearing loss, stress/noise pollution |
| Hazard-16 | Powder | Respiratory tract, lung diseases |
| Hazard-17 | Dump | Microbial diseases, rodents and pests, etc. |
| Hazard-18 | Working with lifting tools | Injuries, death |
| Hazard-19 | Brightness, excessive lighting, and insufficient lighting | Eye conditions, stress |
| Hazard-20 | Energy cut-out (electric, pneumatic, hydraulic) | Burning, distortion, and death |
| Hazard-21 | Exposure to chemical liquid and gas | Cancer, burns, eye conditions, and irritation |
| Hazard-22 | Emergency events | Injury, death |
| Hazard-23 | Professional competence and experience | Injury, death |

### 4.3.2 *Validation Study on the Results*

In this subsection, three validation tests on obtained ranking results are performed. The first validation study concerns a comparison between the results of the existed approach (IVPFVIKOR under Fine–Kinney's method) and classical Fine–Kinney method. We then observe the variations in hazard rankings. The results are shown in Fig. 4.4.

**Table 4.2** Linguistic assessment of the hazards by OHS experts

| Hazard | Probability | | | | | | Exposure | | | | | | Consequence | | | | | |
|---|---|---|---|---|---|---|---|---|---|---|---|---|---|---|---|---|---|---|
| | E-1 | | E-2 | | E-3 | | E-1 | | E-2 | | E-3 | | E-1 | | E-2 | | E-3 | |
| | *u* | *v* | *u* | *v* | *u* | *v* | *u* | *v* | *u* | *v* | *u* | *v* | *u* | *v* | *u* | *v* | *u* | *v* |
| Hazard-1 | 0.65 | 0.35 | 0.65 | 0.35 | 0.65 | 0.35 | 0.75 | 0.25 | 0.85 | 0.15 | 0.75 | 0.25 | 0.35 | 0.65 | 0.35 | 0.65 | 0.35 | 0.65 |
| Hazard-2 | 0.65 | 0.35 | 0.65 | 0.35 | 0.65 | 0.35 | 0.75 | 0.25 | 0.75 | 0.25 | 0.75 | 0.25 | 0.35 | 0.65 | 0.25 | 0.75 | 0.35 | 0.65 |
| Hazard-3 | 0.75 | 0.25 | 0.75 | 0.25 | 0.75 | 0.25 | 0.75 | 0.25 | 0.75 | 0.25 | 0.75 | 0.25 | 0.35 | 0.65 | 0.35 | 0.65 | 0.35 | 0.65 |
| Hazard-4 | 0.65 | 0.35 | 0.65 | 0.35 | 0.65 | 0.35 | 0.75 | 0.25 | 0.75 | 0.25 | 0.85 | 0.15 | 0.35 | 0.65 | 0.35 | 0.65 | 0.25 | 0.75 |
| Hazard-5 | 0.75 | 0.25 | 0.75 | 0.25 | 0.75 | 0.25 | 0.75 | 0.25 | 0.85 | 0.15 | 0.85 | 0.15 | 0.5 | 0.45 | 0.5 | 0.45 | 0.5 | 0.45 |
| Hazard-6 | 0.65 | 0.35 | 0.65 | 0.35 | 0.65 | 0.35 | 0.85 | 0.15 | 0.85 | 0.15 | 0.85 | 0.15 | 0.35 | 0.65 | 0.35 | 0.65 | 0.35 | 0.65 |
| Hazard-7 | 0.65 | 0.35 | 0.65 | 0.35 | 0.65 | 0.35 | 0.65 | 0.35 | 0.65 | 0.35 | 0.65 | 0.35 | 0.35 | 0.65 | 0.35 | 0.65 | 0.35 | 0.65 |
| Hazard-8 | 0.65 | 0.35 | 0.65 | 0.35 | 0.65 | 0.35 | 0.85 | 0.15 | 0.85 | 0.15 | 0.85 | 0.15 | 0.35 | 0.65 | 0.35 | 0.65 | 0.35 | 0.65 |
| Hazard-9 | 0.65 | 0.35 | 0.65 | 0.35 | 0.65 | 0.35 | 0.75 | 0.25 | 0.75 | 0.25 | 0.75 | 0.25 | 0.5 | 0.45 | 0.5 | 0.45 | 0.5 | 0.45 |
| Hazard-10 | 0.5 | 0.45 | 0.65 | 0.35 | 0.5 | 0.45 | 0.25 | 0.75 | 0.35 | 0.65 | 0.35 | 0.65 | 0.65 | 0.35 | 0.65 | 0.35 | 0.65 | 0.35 |
| Hazard-11 | 0.5 | 0.45 | 0.5 | 0.45 | 0.5 | 0.45 | 0.75 | 0.25 | 0.75 | 0.25 | 0.85 | 0.15 | 0.35 | 0.65 | 0.35 | 0.65 | 0.35 | 0.65 |
| Hazard-12 | 0.5 | 0.45 | 0.5 | 0.45 | 0.35 | 0.65 | 0.85 | 0.15 | 0.85 | 0.15 | 0.85 | 0.15 | 0.5 | 0.45 | 0.5 | 0.45 | 0.5 | 0.45 |
| Hazard-13 | 0.5 | 0.45 | 0.5 | 0.45 | 0.5 | 0.45 | 0.35 | 0.65 | 0.35 | 0.65 | 0.35 | 0.65 | 0.15 | 0.85 | 0.15 | 0.85 | 0.15 | 0.85 |
| Hazard-14 | 0.65 | 0.35 | 0.65 | 0.35 | 0.65 | 0.35 | 0.75 | 0.25 | 0.75 | 0.25 | 0.75 | 0.25 | 0.35 | 0.65 | 0.35 | 0.65 | 0.35 | 0.65 |
| Hazard-15 | 0.65 | 0.35 | 0.65 | 0.35 | 0.65 | 0.35 | 0.85 | 0.15 | 0.85 | 0.15 | 0.75 | 0.25 | 0.5 | 0.45 | 0.5 | 0.45 | 0.5 | 0.45 |
| Hazard-16 | 0.5 | 0.45 | 0.5 | 0.45 | 0.35 | 0.65 | 0.35 | 0.65 | 0.25 | 0.75 | 0.35 | 0.65 | 0.35 | 0.65 | 0.35 | 0.65 | 0.35 | 0.65 |
| Hazard-17 | 0.5 | 0.45 | 0.5 | 0.45 | 0.5 | 0.45 | 0.35 | 0.65 | 0.35 | 0.65 | 0.25 | 0.75 | 0.25 | 0.75 | 0.15 | 0.85 | 0.15 | 0.85 |
| Hazard-18 | 0.65 | 0.35 | 0.65 | 0.35 | 0.65 | 0.35 | 0.65 | 0.35 | 0.65 | 0.35 | 0.65 | 0.35 | 0.5 | 0.45 | 0.5 | 0.45 | 0.5 | 0.45 |
| Hazard-19 | 0.65 | 0.35 | 0.65 | 0.35 | 0.65 | 0.35 | 0.35 | 0.65 | 0.35 | 0.65 | 0.35 | 0.65 | 0.5 | 0.45 | 0.5 | 0.45 | 0.5 | 0.45 |

(continued)

**Table 4.2** (continued)

| Hazard | Probability | | | | | | Exposure | | | | | | Consequence | | | | | |
|---|---|---|---|---|---|---|---|---|---|---|---|---|---|---|---|---|---|---|
| | E-1 | | E-2 | | E-3 | | E-1 | | E-2 | | E-3 | | E-1 | | E-2 | | E-3 | |
| | $u$ | $v$ | $u$ | $v$ | $u$ | $v$ | $u$ | $v$ | $u$ | $v$ | $u$ | $v$ | $u$ | $v$ | $u$ | $v$ | $u$ | $v$ |
| Hazard-20 | 0.5 | 0.45 | 0.5 | 0.45 | 0.5 | 0.45 | 0.75 | 0.25 | 0.75 | 0.25 | 0.75 | 0.25 | 0.65 | 0.35 | 0.65 | 0.35 | 0.65 | 0.35 |
| Hazard-21 | 0.65 | 0.35 | 0.65 | 0.35 | 0.65 | 0.35 | 0.75 | 0.25 | 0.75 | 0.25 | 0.75 | 0.25 | 0.5 | 0.45 | 0.5 | 0.45 | 0.5 | 0.45 |
| Hazard-22 | 0.5 | 0.45 | 0.5 | 0.45 | 0.5 | 0.45 | 0.65 | 0.35 | 0.65 | 0.35 | 0.65 | 0.35 | 0.5 | 0.45 | 0.5 | 0.45 | 0.5 | 0.45 |
| Hazard-23 | 0.65 | 0.35 | 0.65 | 0.35 | 0.75 | 0.25 | 0.75 | 0.25 | 0.75 | 0.25 | 0.75 | 0.25 | 0.5 | 0.45 | 0.5 | 0.45 | 0.5 | 0.45 |

**Table 4.3** The Pythagorean fuzzy decision matrix (aggregated)

| Hazard | Probability | | Exposure | | Consequence | |
|---|---|---|---|---|---|---|
| | *u* | *v* | *u* | *v* | *u* | *v* |
| Hazard-1 | 0.650 | 0.350 | 0.786 | 0.214 | 0.350 | 0.650 |
| Hazard-2 | 0.650 | 0.350 | 0.750 | 0.250 | 0.316 | 0.679 |
| Hazard-3 | 0.750 | 0.250 | 0.750 | 0.250 | 0.350 | 0.650 |
| Hazard-4 | 0.650 | 0.350 | 0.786 | 0.214 | 0.316 | 0.679 |
| Hazard-5 | 0.750 | 0.250 | 0.817 | 0.184 | 0.500 | 0.450 |
| Hazard-6 | 0.650 | 0.350 | 0.850 | 0.150 | 0.350 | 0.650 |
| Hazard-7 | 0.650 | 0.350 | 0.650 | 0.350 | 0.350 | 0.650 |
| Hazard-8 | 0.650 | 0.350 | 0.850 | 0.150 | 0.350 | 0.650 |
| Hazard-9 | 0.650 | 0.350 | 0.750 | 0.250 | 0.500 | 0.450 |
| Hazard-10 | 0.554 | 0.417 | 0.315 | 0.688 | 0.650 | 0.350 |
| Hazard-11 | 0.500 | 0.450 | 0.786 | 0.214 | 0.350 | 0.650 |
| Hazard-12 | 0.462 | 0.502 | 0.850 | 0.150 | 0.500 | 0.450 |
| Hazard-13 | 0.500 | 0.450 | 0.350 | 0.650 | 0.150 | 0.850 |
| Hazard-14 | 0.650 | 0.350 | 0.750 | 0.250 | 0.350 | 0.650 |
| Hazard-15 | 0.650 | 0.350 | 0.826 | 0.175 | 0.500 | 0.450 |
| Hazard-16 | 0.462 | 0.502 | 0.324 | 0.679 | 0.350 | 0.650 |
| Hazard-17 | 0.500 | 0.450 | 0.324 | 0.679 | 0.184 | 0.808 |
| Hazard-18 | 0.650 | 0.350 | 0.650 | 0.350 | 0.500 | 0.450 |
| Hazard-19 | 0.650 | 0.350 | 0.350 | 0.650 | 0.500 | 0.450 |
| Hazard-20 | 0.500 | 0.450 | 0.750 | 0.250 | 0.650 | 0.350 |
| Hazard-21 | 0.650 | 0.350 | 0.750 | 0.250 | 0.500 | 0.450 |
| Hazard-22 | 0.500 | 0.450 | 0.650 | 0.350 | 0.500 | 0.450 |
| Hazard-23 | 0.685 | 0.316 | 0.750 | 0.250 | 0.500 | 0.450 |

It is observed from Fig. 4.4 that by both approaches, *Hazard-5* is ranked as the most critical hazard, followed by *Hazard-15*. It has also seen that the least important three hazards (*Hazard-13*, *Hazard-16*, and *Hazard-17*) have partially the same ranking according to both approaches. When we compare the results obtained by both approaches, we observe that there are very small rank variations between them. The Spearman rank correlation between the two approaches is obtained as 0.928. That means there exists a high correlation between the ranking orders of two approaches so that it can be claimed that this proposed approach is applicable for occupational risk assessment in the Fine–Kinney domain.

As a second validation study, we analyze the difference in the ranking of hazards in times of changing of Fine–Kinney parameters' weights. Therefore, we apply four different weight vectors. The weight vectors of Fine–Kinney parameters designed for the sensitivity analysis are given in Table 4.5. The ranking of hazards with respect to four different weight vectors is shown in Table 4.6.

**Table 4.4** $S$, $R$, and $Q$ values and ranking orders for each hazard

| Hazard | $Q_a$ value | Ranking |
|---|---|---|
| Hazard-1 | 0.289 | 8 |
| Hazard-2 | 0.366 | 13 |
| Hazard-3 | 0.234 | 6 |
| Hazard-4 | 0.349 | 12 |
| Hazard-5 | 0.000 | 1 |
| Hazard-6 | 0.259 | 7 |
| Hazard-7 | 0.346 | 10 |
| Hazard-8 | 0.259 | 7 |
| Hazard-9 | 0.102 | 4 |
| Hazard-10 | 0.517 | 18 |
| Hazard-11 | 0.461 | 15 |
| Hazard-12 | 0.445 | 14 |
| Hazard-13 | 1.000 | 21 |
| Hazard-14 | 0.305 | 9 |
| Hazard-15 | 0.067 | 2 |
| Hazard-16 | 0.685 | 19 |
| Hazard-17 | 0.895 | 20 |
| Hazard-18 | 0.180 | 5 |
| Hazard-19 | 0.506 | 17 |
| Hazard-20 | 0.349 | 11 |
| Hazard-21 | 0.102 | 4 |
| Hazard-22 | 0.463 | 16 |
| Hazard-23 | 0.078 | 3 |

It can be observed from Table 4.6 that when the weights change, there exist variations in the ranking of hazards. Therefore, our proposed approach is sensitive to Fine–Kinney risk parameters' weights. *Hazard-5* is mostly ranked as the most critical hazard according to all the weight vectors. There is no change in the ranking of *Hazard-16* for all combinations. It lies in the 19th place in the ranking. When compared to the results with the ones similar to this study from the literature, we can say that the ranking result obtained by our proposed approach is credible and applicable.

We also calculated the Spearman rank correlation (RHO) between weight vectors by an online calculator. The obtained results are given in Table 4.7. Results show that there exist high correlations between the ranking orders obtained by four different weight vectors. Since all values are close to 1. To this end, it can be claimed that this proposed approach is sensitive to the changing of the weight values. It is an expected output when considering similar attempts from the literature.

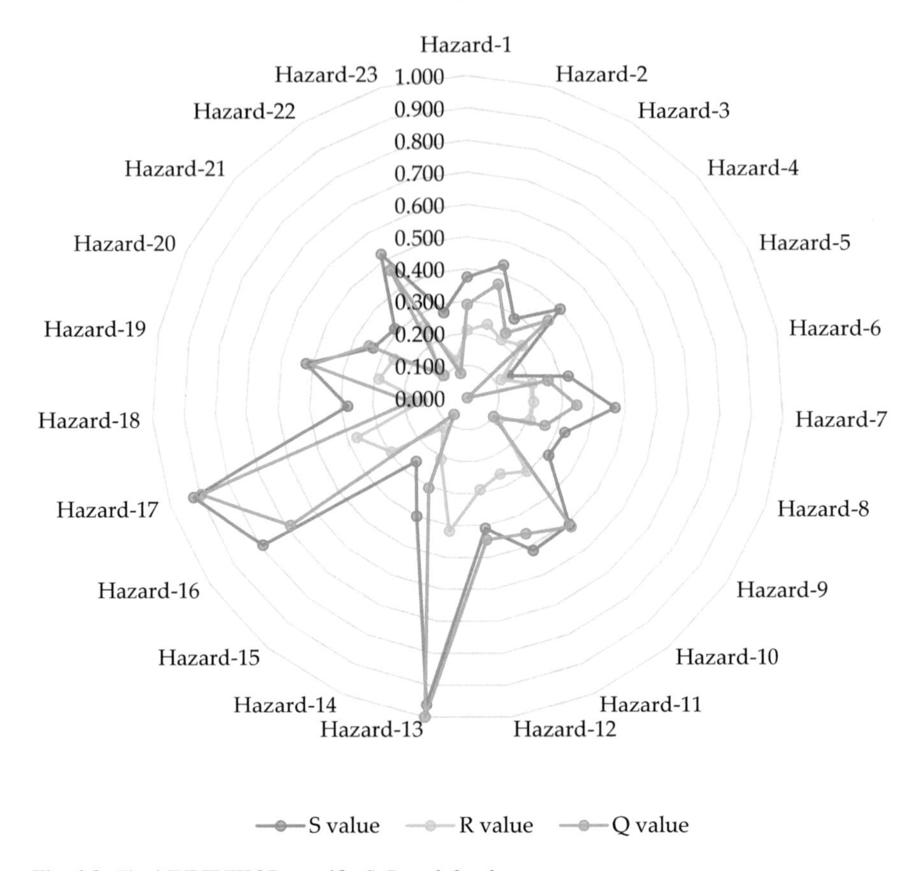


**Fig. 4.3** Final IVPFVIKOR-specific *S*, *R*, and *Q* values

A third validation study (as a sensitivity analysis) was performed in the results of the proposed approach by varying value of $v$ (maximum group utility) which is set to 0.5. Ten different scenarios (excluding the experiment of $v = 0.5$) are tried to observe the variability. Results of Q values from this sensitivity analysis are given in Fig. 4.5. It is clearly understood from Fig. 4.4 that *Hazard-5* has the best ranking for each case, *Hazard-13* has the worst ranking for each $v$ value changing situation.

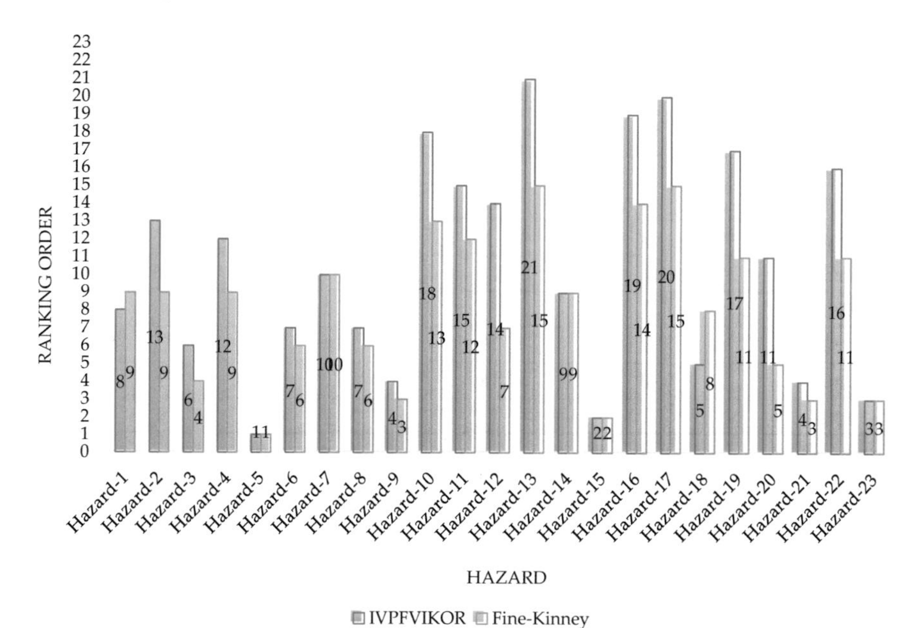


**Fig. 4.4** Comparison of rankings by proposed and classic approaches

**Table 4.5** The weight vectors designed for the sensitivity analysis

| Weight vector | Parameter | Weight value |
|---|---|---|
| Weight vector-1 ($W_1$) | Probability | 0.289 |
| | Exposure | 0.293 |
| | Consequence | 0.418 |
| Weight vector-2 ($W_2$) | Probability | 0.250 |
| | Exposure | 0.200 |
| | Consequence | 0.550 |
| Weight vector-3 ($W_3$) | Probability | 0.333 |
| | Exposure | 0.333 |
| | Consequence | 0.333 |
| Weight vector-4 ($W_4$) | Probability | 0.400 |
| | Exposure | 0.400 |
| | Consequence | 0.200 |

**Table 4.6** Ranking order changes in times of parameters' weight changes

| Hazard | Ranking order of the hazard | | | |
|---|---|---|---|---|
| | $W_1$ | $W_2$ | $W_3$ | $W_4$ |
| Hazard-1 | 8 | 13 | 7 | 7 |
| Hazard-2 | 13 | 17 | 12 | 10 |
| Hazard-3 | 6 | 9 | 5 | 2 |
| Hazard-4 | 12 | 16 | 10 | 8 |
| Hazard-5 | 1 | 1 | 1 | 1 |
| Hazard-6 | 7 | 11 | 6 | 5 |
| Hazard-7 | 10 | 15 | 11 | 12 |
| Hazard-8 | 7 | 11 | 6 | 5 |
| Hazard-9 | 4 | 4 | 4 | 6 |
| Hazard-10 | 18 | 7 | 18 | 18 |
| Hazard-11 | 15 | 18 | 14 | 14 |
| Hazard-12 | 14 | 10 | 15 | 15 |
| Hazard-13 | 21 | 21 | 21 | 20 |
| Hazard-14 | 9 | 14 | 9 | 9 |
| Hazard-15 | 2 | 2 | 3 | 4 |
| Hazard-16 | 19 | 19 | 19 | 19 |
| Hazard-17 | 20 | 20 | 20 | 21 |
| Hazard-18 | 5 | 5 | 8 | 11 |
| Hazard-19 | 17 | 8 | 17 | 17 |
| Hazard-20 | 11 | 6 | 13 | 13 |
| Hazard-21 | 4 | 4 | 4 | 6 |
| Hazard-22 | 16 | 12 | 16 | 16 |
| Hazard-23 | 3 | 3 | 2 | 3 |

**Table 4.7** Results of Spearman's RHO between weight vectors

| | $W_1$ | $W_2$ | $W_3$ | $W_4$ |
|---|---|---|---|---|
| $W_1$ | – | 0.763 | 0.983 | 0.925 |
| $W_2$ | | – | 0.704 | 0.607 |
| $W_3$ | | | – | 0.968 |
| $W_4$ | | | | – |

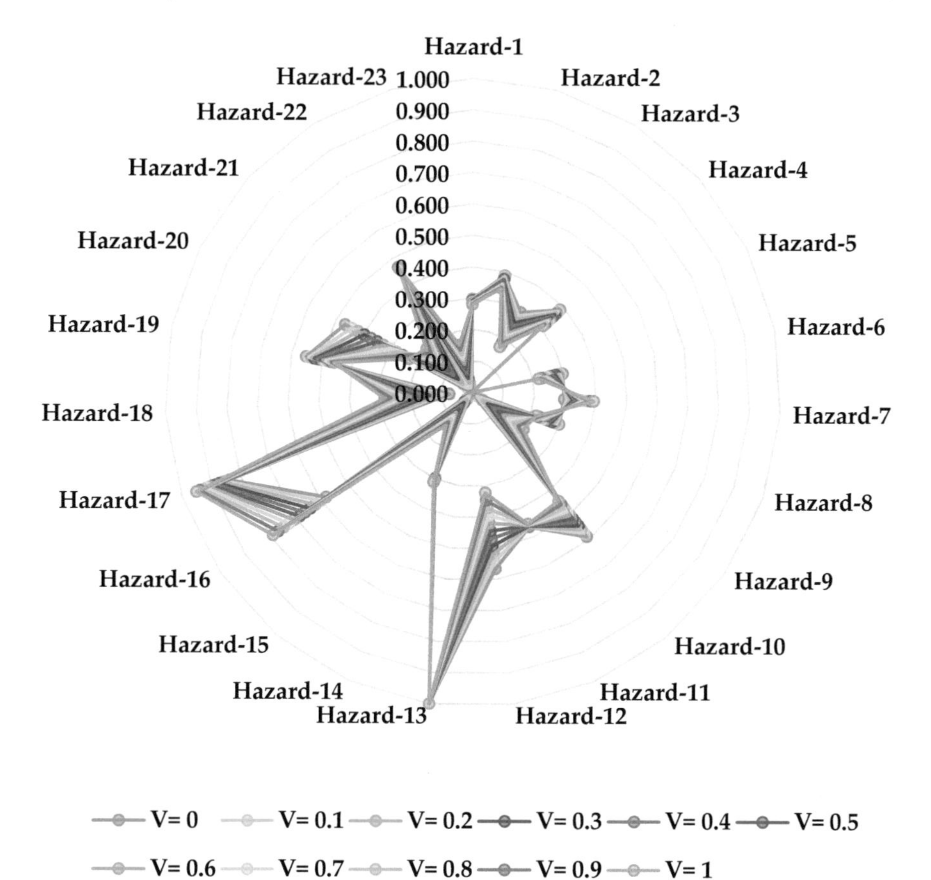


**Fig. 4.5** IVPFVIKOR Q values according to different $v$ values

## 4.4 Python Implementation of the Proposed Approach

```
# Chapter 4
# import required libraries
import numpy as np

n_criteria = 3  # P, E, C
ns_value = 2  # μ, v
n_expert = 3  # DM-1,DM-2, DM-3
n_hazard = 23
v = 0.5
# Probability (p), Exposure(e), Consequence(c)
pec_weight = [0.289, 0.293, 0.418]
expert_weight = [0.4, 0.3, 0.3]

linguistic_term = [
    # Linguistic term, u, v
    ["VL", 0.15, 0.85],
    ["L", 0.25, 0.75],
    ["ML", 0.35, 0.65],
    ["M", 0.5, 0.45],
    ["MH", 0.65, 0.35],
    ["H", 0.75, 0.25],
    ["VH", 0.85, 0.15]
]
```

```
# ehe1: 1st expert's hazard evaluation
ehe1 = [
    # Hazard id,   P, E, C,
    ["FM1", "MH", "H", "ML"],
    ["FM2", "MH", "H", "ML"],
    ["FM3", "H", "H", "ML"],
    ["FM4", "MH", "H", "ML"],
    ["FM5", "H", "H", "M"],
    ["FM6", "MH", "VH", "ML"],
    ["FM7", "MH", "MH", "ML"],
    ["FM8", "MH", "VH", "ML"],
    ["FM9", "MH", "H", "M"],
    ["FM10", "M", "L", "MH"],
    ["FM11", "M", "H", "ML"],
    ["FM12", "M", "VH", "M"],
    ["FM13", "M", "ML", "VL"],
    ["FM14", "MH", "H", "ML"],
    ["FM15", "MH", "VH", "M"],
    ["FM16", "M", "ML", "ML"],
    ["FM17", "M", "ML", "L"],
    ["FM18", "MH", "MH", "M"],
    ["FM19", "MH", "ML", "M"],
    ["FM20", "M", "H", "MH"],
    ["FM21", "MH", "H", "M"],
    ["FM22", "M", "MH", "M"],
    ["FM23", "MH", "H", "M"]
]
# ehe2: 2nd expert's hazard evaluation
```

```
ehe2 = [
  # Hazard id,   P, E, C,
  ["FM1", "MH", "VH", "ML"],
  ["FM2", "MH", "H", "L"],
  ["FM3", "H", "H", "ML"],
  ["FM4", "MH", "H", "ML"],
  ["FM5", "H", "VH", "M"],
  ["FM6", "MH", "VH", "ML"],
  ["FM7", "MH", "MH", "ML"],
  ["FM8", "MH", "VH", "ML"],
  ["FM9", "MH", "H", "M"],
  ["FM10", "MH", "ML", "MH"],
  ["FM11", "M", "H", "ML"],
  ["FM12", "M", "VH", "M"],
  ["FM13", "M", "ML", "VL"],
  ["FM14", "MH", "H", "ML"],
  ["FM15", "MH", "VH", "M"],
  ["FM16", "M", "L", "ML"],
  ["FM17", "M", "ML", "VL"],
  ["FM18", "MH", "MH", "M"],
  ["FM19", "MH", "ML", "M"],
  ["FM20", "M", "H", "MH"],
  ["FM21", "MH", "H", "M"],
  ["FM22", "M", "MH", "M"],
  ["FM23", "MH", "H", "M"]
]
# ehe3: 3rd expert's hazard evaluation
```

```
ehe3 = [
  # Hazard id,   P, E, C,
  ["FM1", "MH", "H", "ML"],
  ["FM2", "MH", "H", "ML"],
  ["FM3", "H", "H", "ML"],
  ["FM4", "MH", "VH", "L"],
  ["FM5", "H", "VH", "M"],
  ["FM6", "MH", "VH", "ML"],
  ["FM7", "MH", "MH", "ML"],
  ["FM8", "MH", "VH", "ML"],
  ["FM9", "MH", "H", "M"],
  ["FM10", "M", "ML", "MH"],
  ["FM11", "M", "VH", "ML"],
  ["FM12", "ML", "VH", "M"],
  ["FM13", "M", "ML", "VL"],
  ["FM14", "MH", "H", "ML"],
  ["FM15", "MH", "H", "M"],
  ["FM16", "ML", "ML", "ML"],
  ["FM17", "M", "L", "VL"],
  ["FM18", "MH", "MH", "M"],
  ["FM19", "MH", "ML", "M"],
  ["FM20", "M", "H", "MH"],
  ["FM21", "MH", "H", "M"],
  ["FM22", "M", "MH", "M"],
  ["FM23", "H", "H", "M"]
]
```

```
def rank(vector, da): # da -1:descending, 1:ascending
    order = np.zeros([len(vector), 1])
    unique_val = da * np.sort(da * np.unique(vector))
    for ix in range(0, len(unique_val)):
        order[np.argwhere(vector == unique_val[ix])] = ix + 1
    return order

def print_result(order, vector):
    print('Hazard Id, Rank, Value')
    for ix in range(0, len(order)):
        print(ehe1[ix][0], ', ', int(order[ix]), ', ', vector[ix])

ehe = np.asarray([ehe1, ehe2, ehe3])
ehe_numeric = np.zeros([n_expert, n_hazard, n_criteria, ns_value], dtype=float)
for i in range(0, n_expert):
    for j in range(0, n_hazard):
        for k in range(0, n_criteria):
            for lt in linguistic_term:
                if ehe[i][j][k + 1] == lt[0]:
                    ehe_numeric[i][j][k] = lt[1:]

ehe_numeric[:, :, 0:2, 0] = 1 - np.square(ehe_numeric[:, :, 0:2, 0])
aggregated_matrix = np.zeros([n_hazard, n_criteria, ns_value], dtype=float)
for j in range(0, n_hazard):
    for k in range(0, n_criteria):
        for l in range(0, ns_value):
            val = 1
            for i in range(0, n_expert):
                val *= np.power(ehe_numeric[i][j][k][l], expert_weight[i])
            if (k == 0 or k == 1) and l == 0:
                val = np.sqrt(1 - val)
            aggregated_matrix[j][k][l] = val
```

```
pi = np.zeros([n_hazard, n_criteria], dtype=float)
sP = np.zeros([n_hazard, n_criteria], dtype=float)
for i in range(0, n_hazard):
    for j in range(0, n_criteria):
        pi[i][j] = np.sqrt(1 - np.square(aggregated_matrix[i][j][0])
                        - np.square(aggregated_matrix[i][j][1]))
        sP[i][j] = (1 + np.square(aggregated_matrix[i][j][0])
                  - np.square(aggregated_matrix[i][j][1])) / 2

fa_positive = np.zeros([n_criteria, ns_value])
fa_negative = np.zeros([n_criteria, ns_value])
for i in range(0, n_criteria):
    fa_positive[i][:] = aggregated_matrix[np.argmax(sP[:, i]), i, :]
    fa_negative[i][:] = aggregated_matrix[np.argmin(sP[:, i]), i, :]

fa_pi = np.zeros([2, n_criteria], dtype=float)
fa_pi[0][:] = np.sqrt(1 - np.sum(np.square(fa_positive), axis=1))
fa_pi[1][:] = np.sqrt(1 - np.sum(np.square(fa_negative), axis=1))

x = np.sum(np.abs(np.square(aggregated_matrix)
           - np.square(fa_positive)), axis=2)
y = np.abs(np.square(pi) - np.square(fa_pi[0][:]))
z = (x + y) / 2

a = np.sum(np.abs(np.square(fa_positive)
           - np.square(fa_negative)), axis=1)
b = np.abs(np.square(fa_pi[0][:]) - np.square(fa_pi[1][:]))
c = (a + b) / 2

pec_matrix = z / c

rank_s = pec_matrix.dot(pec_weight)
rank_r = np.max(pec_matrix * pec_weight, axis=1)
s_positive, s_negative = np.min(rank_s), np.max(rank_s)
r_positive, r_negative = np.min(rank_r), np.max(rank_r)

delta_s = s_negative - s_positive
delta_r = r_negative - r_positive
```

```
delta_rank_s = v * (rank_s - s_positive) / delta_s
delta_rank_r = (1 - v) * (rank_r - r_positive) / delta_r

rank_q = delta_rank_s + delta_rank_r

hazard_rank = rank(rank_q, 1)
print_result(hazard_rank, rank_q)
'''
Output:
Hazard Id, Rank, Value
FM1 , 8 , 0.28924089873674574
FM2 , 13 , 0.36592427613254086
FM3 , 6 , 0.23399873081233355
FM4 , 12 , 0.34977827261006933
FM5 , 1 , 0.0
FM6 , 7 , 0.2591328247185968
FM7 , 10 , 0.3458592201072608
FM8 , 7 , 0.2591328247185968
FM9 , 4 , 0.1016189696964118
FM10 , 18 , 0.5109512950538306
FM11 , 15 , 0.461175544634636
FM12 , 14 , 0.44519934021149643
FM13 , 21 , 1.0
FM14 , 9 , 0.3053869022592176
FM15 , 2 , 0.06710479476008396
FM16 , 19 , 0.6804104416122312
FM17 , 20 , 0.8950645125167197
FM18 , 5 , 0.17700729795659575
FM19 , 17 , 0.5004040897223132
FM20 , 11 , 0.34874478216825633
FM21 , 4 , 0.1016189696964118
FM22 , 16 , 0.4631487404598892
FM23 , 3 , 0.0780858225637129
'''
```

## References

1. Yager, R. R. (2014). Pythagorean membership grades in multicriteria decision making. *IEEE Transactions on Fuzzy Systems, 22*(4), 958–965.
2. Gul, M. (2018). Application of Pythagorean fuzzy AHP and VIKOR methods in occupational health and safety risk assessment: The case of a gun and rifle barrel external surface oxidation and colouring unit. *International journal of occupational safety and ergonomics*. https://doi.org/10.1080/10803548.2018.1492251.
3. Gul, M., & Ak, M. F. (2018). A comparative outline for quantifying risk ratings in occupational health and safety risk assessment. *Journal of Cleaner Production, 196,* 653–664.
4. Gul, M., Guneri, A. F., & Nasirli, S. M. (2018). A fuzzy-based model for risk assessment of routes in oil transportation. *International Journal of Environmental Science and Technology*, 1–16.
5. Ilbahar, E., Karasan, A., Cebi, S., & Kahraman, C. (2018). A novel approach to risk assessment for occupational health and safety using Pythagorean fuzzy AHP & fuzzy inference system. *Safety Science, 103,* 124–136.
6. Karasan, A., Ilbahar, E., Cebi, S., & Kahraman, C. (2018). A new risk assessment approach: Safety and Critical Effect Analysis (SCEA) and its extension with Pythagorean fuzzy sets. *Safety Science, 108,* 173–187.
7. Mete, S., Serin, F., Oz, N. E., & Gul, M. (2019). A decision-support system based on Pythagorean fuzzy VIKOR for occupational risk assessment of a natural gas pipeline construction. *Journal of Natural Gas Science and Engineering, 71,* 102979.
8. Oz, N. E., Mete, S., Serin, F., & Gul, M. (2019). Risk assessment for clearing and grading process of a natural gas pipeline project: An extended TOPSIS model with Pythagorean fuzzy sets for prioritizing hazards. *Human and Ecological Risk Assessment: An International Journal, 25*(6), 1615–1632.
9. Ak, M. F., & Gul, M. (2019). AHP–TOPSIS integration extended with Pythagorean fuzzy sets for information security risk analysis. *Complex & Intelligent Systems, 5*(2), 113–126.
10. Gul, M., Ak, M. F., & Guneri, A. F. (2019). Pythagorean fuzzy VIKOR-based approach for safety risk assessment in mine industry. *Journal of Safety Research, 69,* 135–153.
11. Yucesan, M., & Gul, M. (2019). Hospital service quality evaluation: an integrated model based on Pythagorean fuzzy AHP and fuzzy TOPSIS. *Soft Computing*, 1–19.
12. Ozdemir, Y., & Gul, M. (2019). Measuring development levels of NUTS-2 regions in Turkey based on capabilities approach and multi-criteria decision-making. *Computers & Industrial Engineering, 128,* 150–169.
13. Mete, S. (2019). Assessing occupational risks in pipeline construction using FMEA-based AHP-MOORA integrated approach under Pythagorean fuzzy environment. *Human and Ecological Risk Assessment: An International Journal, 25*(7), 1645–1660.
14. Yucesan, M., & Kahraman, G. (2019). Risk evaluation and prevention in hydropower plant operations: A model based on Pythagorean fuzzy AHP. *Energy policy, 126,* 343–351.
15. Zeng, S., Chen, J., & Li, X. (2016). A hybrid method for pythagorean fuzzy multiple-criteria decision making. *International Journal of Information Technology & Decision Making, 15*(02), 403–422.
16. Zhang, X., & Xu, Z. (2014). Extension of TOPSIS to multiple criteria decision making with Pythagorean fuzzy sets. *International Journal of Intelligent Systems, 29*(12), 1061–1078.
17. Opricovic, S. (1998). *Multicriteria optimization of civil engineering systems*. Belgrade: Faculty of Civil Engineering.
18. Gul, M., Celik, E., Aydin, N., Gumus, A. T., & Guneri, A. F. (2016). A state of the art literature review of VIKOR and its fuzzy extensions on applications. *Applied Soft Computing, 46,* 60–89.
19. Peng, X., & Selvachandran, G. (2019). Pythagorean fuzzy set: state of the art and future directions. *Artificial Intelligence Review, 52*(3), 1873–1927.
20. Yu, C., Shao, Y., Wang, K., & Zhang, L. (2019). A group decision making sustainable supplier selection approach using extended TOPSIS under interval-valued Pythagorean fuzzy environment. *Expert Systems with Applications, 121,* 1–17.

21. Garg, H. (2016). A novel accuracy function under interval-valued Pythagorean fuzzy environment for solving multicriteria decision making problem. *Journal of Intelligent & Fuzzy Systems, 31*(1), 529–540.
22. Cui, F. B., You, X. Y., Shi, H., & Liu, H. C. (2018). Optimal siting of electric vehicle charging stations using Pythagorean fuzzy VIKOR approach. *Mathematical Problems in Engineering.*
23. Yazdi, M. (2018). Footprint of knowledge acquisition improvement in failure diagnosis analysis. *Quality and Reliability Engineering International.* https://doi.org/10.1002/qre.2408.
24. Kabir, S., Yazdi, M., Aizpurua, J. I., & Papadopoulos, Y. (2018). Uncertainty-aware dynamic reliability analysis framework for complex systems. *IEEE Access.* https://doi.org/10.1109/ACCESS.2018.2843166.
25. Gul, M., Guven, B., & Guneri, A. F. (2018). A new Fine-Kinney-based risk assessment framework using FAHP-FVIKOR incorporation. *Journal of Loss Prevention in the Process Industries, 53,* 3–16.

# Chapter 5
# Fine–Kinney-Based Occupational Risk Assessment Using Intuitionistic Fuzzy TODIM

**Abstract** This chapter applies a novel occupational risk assessment approach which merges TODIM with the Fine–Kinney method under the intuitionistic fuzzy set concept. Risk parameters of Fine–Kinney and OHS experts are weighted by an intuitionistic fuzzy weighted averaging (IFWA) aggregation operator. Hence, hazards are quantitatively evaluated and prioritized using the proposed approach. To illustrate the novel risk assessment approach, processes of the gun and rifle assembly line of a factory are handled. A comprehensive risk assessment is carried out to improve operational safety and reliability in the industry. We adapt intuitionistic fuzzy sets in the existing study since they reflect uncertainty with the aid of their membership and nonmembership functions in decision-making better than classical fuzzy extensions. An additional sensitivity analysis by changing the attenuation parameter of TODIM is performed to test the validity of the approach. Finally, the Python codes in the implementation of the proposed approach are given for scholars and practitioners for usage in further studies.

## 5.1 Intuitionistic Fuzzy Sets and TODIM

In this part, prior to explaining the proposed approach, some preliminaries are provided regarding intuitionistic fuzzy sets. Then, the algorithm intuitionistic fuzzy TODIM (IFTODIM) is presented in detail.

### *5.1.1 General View on Intuitionistic Fuzzy Sets*

Intuitionistic fuzzy sets have been initially suggested by Atanassov [1] and have been applied by many scholars to various areas to handle uncertainty. These sets are defined as membership and nonmembership functions. In intuitionistic fuzzy sets, the degrees of membership and nonmembership are equal to 1. These sets are combined with various MCDM methods such as AHP [2, 3], ANP [4], TOPSIS [5], VIKOR [6],



M. Gul et al., *Fine–Kinney-Based Fuzzy Multi-criteria Occupational Risk Assessment*,
Studies in Fuzziness and Soft Computing 398,
https://doi.org/10.1007/978-3-030-52148-6_5

PROMETHEE [7], ELECTRE [8], DEMATEL [9] and as well as TODIM [10–12]. In the following, some definitions of intuitionistic fuzzy sets are given.

**Definition 1** Let X be a set in a universe of discourse. An intuitionistic fuzzy set I has the form in the following [10–12]:

$$I = \{< x,\ I(\mu_I(x), v_I(x)) > | x \in X\} \tag{5.1}$$

where $\mu_I(x) : X \mapsto [0, 1]$ describes the degree of membership and $v_I(x) : X \mapsto [0, 1]$ describes the degree of nonmembership of the element $x \in X$ to $I$, separately, and, for each $x \in X$, it takes

$$0 \leq \mu_I(x) + v_I(x) \leq 1 \tag{5.2}$$

**Definition 2** Let $I_1 = I(\mu_{I_1}, v_{I_1})$ and $I_2 = I(\mu_{I_2}, v_{I_2})$ be two intuitionistic fuzzy numbers, and $\lambda > 0$, then the operations on these two intuitionistic fuzzy numbers are described as below [10–12]:

$$I_1 \oplus I_2 = I(\mu_{I_1} + \mu_{I_2} - \mu_{I_1}\mu_{I_2},\ v_{I_1}v_{I_2}) \tag{5.3}$$

$$I_1 \otimes I_2 = I(\mu_{I_1}\mu_{I_2},\ v_{I_1} + v_{I_2} - v_{I_1}v_{I_2}) \tag{5.4}$$

$$\lambda I_1 = I(1 - (1 - \mu_{I_1}^2)^\lambda, (v_{I_1})^\lambda),\ \lambda > 0 \tag{5.5}$$

$$I_1^\lambda = I((\mu_{I_1})^\lambda, 1 - (1 - v_{I_1}^2)^\lambda),\ \lambda > 0 \tag{5.6}$$

The literature between two intuitionistic fuzzy numbers can be analyzed in detail for more definitions on score functions and ranking procedures [1–12].

### *5.1.2 TODIM Method*

The TODIM (Interactive and Multiple Attribute Decision Making, acronym in Portuguese) approach was presented by Gomes and Lima [13]. It is a multi-criteria decision analysis approach based on prospect theory [14]. In this approach, once each alternative dominance degree considering the decision-maker's behavioral characteristic against others is computed, then, ranking is assigned to each alternative after calculating each alternative overall score [15]. The decision-maker's behavior is considered, and it is the main advantage of the TODIM approach [16]. The procedural steps of the traditional TODIM approach are given in the following.

**Step 1**: The decision matrix is created as in Eq. (5.7). In this chapter, $m$ refers to *hazard* $H_i (i = 1, 2, \ldots, m)$ regarding *n Fine–Kinney risk parameters* $RP_n (j = 1, 2, \ldots, n)$. $X = [x_{ij}]_{mxn}$ is normalized into $Y = [y_{ij}]_{mxn}$ by a normalization formula given in Eqs. (5.8)–(5.9) where $x_{ij} = crisp\,number,\ i \in H,\ j \in RP$.

$$X = \begin{pmatrix} & RP_1 & RP_2 & \ldots & RP_n \\ H_1 & x_{11} & x_{12} & \ldots & x_{1n} \\ H_2 & x_{21} & x_{22} & \ldots & x_{2n} \\ \vdots & \vdots & \vdots & \vdots & \vdots \\ H_m & x_{m1} & x_{m2} & \ldots & x_{mn} \end{pmatrix} \tag{5.7}$$

$$r_{ij} = \frac{x_{ij} - \min(x_{ij})}{\max(x_{ij}) - \min(x_{ij})}$$
$$(for\ benefit\ criterion,\ i = 1, 2, \ldots, m \text{ and } j = 1, 2, \ldots, n) \tag{5.8}$$

$$r_{ij} = \frac{\max(x_{ij}) - x_{ij}}{\max(x_{ij}) - \min(x_{ij})}$$
$$(for\ \cos t\ criterion,\ i = 1, 2, \ldots, m \text{ and } j = 1, 2, \ldots, n) \tag{5.9}$$

**Step 2**: The relative weight $w_{jr}$ of the risk parameter $RP_j$ to the reference risk parameter $RP_r$ is measured by Eq. (5.10).

$$w_{jr} = \frac{w_j}{w_r},\ j = 1, 2, \ldots, n \tag{5.10}$$

where $w_r = \max\{w_j | j = 1, 2, \ldots, n\}$.

**Step 3**: The hazard dominance $H_i$ over hazard $H_p$ according to the risk parameter $RP_j$ is obtained using Eq. (5.11).

$$\phi_j(H_i, H_p) = \begin{cases} \sqrt{\frac{w_{jr}}{\sum_{j=1}^{n} w_{jr}} d(x_{ij}, x_{pj})} & \text{if } x_{ij} > x_{pj} \\ 0 & \text{if } x_{ij} = x_{pj} \\ -\frac{1}{\theta}\sqrt{\frac{\sum_{j=1}^{n} w_{jr}}{w_{jr}} d(x_{ij}, x_{pj})} & \text{if } x_{ij} < x_{pj} \end{cases} \tag{5.11}$$

where $\theta$ = losses attenuation factor.

**Step 4**: The overall dominance of hazard $H_i$ over hazard $H_p$ with respect to the risk parameter $RP_j$ is obtained using Eq. (5.12).

$$\delta_j(H_i, H_p) = \sum_{j=1}^{n} \phi_j(H_i, H_p) \tag{5.12}$$

**Step 5**: The risk degree of hazard $H_i$ is calculated using the following Eq. (5.13). The higher the $\xi(H_i)$, the most serious the risk. Thus, by arranging values of each hazard alternative in descending order, the ranking orders can be found.

$$\xi(H_i)=\frac{\sum_{p=1}^{m}\delta(H_i,H_p)-\min_i\sum_{p=1}^{m}\delta(H_i,H_p)}{\max_i\sum_{p=1}^{m}\delta(H_i,H_p)-\min_i\sum_{p=1}^{m}\delta(H_i,H_p)} \tag{5.13}$$

## 5.2 Proposed Fine–Kinney-Based Approach Using IFTODIM

In this section, we propose a new Fine–Kinney-based occupational risk assessment approach using IFTODIM. In this approach, the three risk parameters for risk assessment are derived from Fine–Kinney. Then, they are weighted using the related steps with the aid of IFTODIM. Initially, linguistic assessments for risk parameter weights and hazards by OHS experts are obtained and aggregated. The weights are also assigned to OHS experts considering a formula which depends on their experience levels. The steps of the proposed Fine–Kinney-based approach using IFTODIM are provided in detail in the following.

**Step 1**: The first phase concerns with the determination of the alternatives (hazards), criteria (Fine–Kinney risk parameters, and OHS experts. The problem in this chapter has $K$ OHS experts $E_k(k = 1\,to\,K)$, $m$ hazards $H_i(i = 1\,to\,m)$, and $n$ Fine–Kinney risk parameters $RP_j(j = 1\,to\,n)$. Each OHS expert $E_k$ has a weight value ($\lambda_k > 0\,and\,\sum\lambda_k = 1$.

**Step 2**: This phase calculates the weights of OHS experts. We follow the procedure of Ref. [5] in weighting the OHS experts as in Eq. (5.14).

$$\lambda_k = \left[\mu_k + \pi_k\left(\frac{\mu_k}{1-\pi_k}\right)\right] / \sum_{k=1}^{K}\left[\mu_k + \pi_k\left(\frac{\mu_k}{1-\pi_k}\right)\right] \tag{5.14}$$

where $\mu_k$ and $\pi_k$ show the membership and nonmembership degree of the corresponding intuitionistic fuzzy numbers.

**Step 3**: In this phase, the decision matrices for each OHS expert in linguistic terms are converted into the intuitionistic fuzzy numbers using the scale in Table 5.1.

**Step 4**: In this phase, the individual decision matrices in Step 3 are aggregated into one final matrix called “aggregated decision matrix” using intuitionistic fuzzy weighted averaging (IFWA) aggregation operator. The formula of the IFWA operator is given in Eq. (5.15).

**Table 5.1** Linguistic scale in intuitionistic fuzzy numbers. Reprinted from Ref. [10], with kind permission from Springer Science+Business Media

| Linguistic term | μ | ν | π |
|---|---|---|---|
| Very Good (VG) | 1.00 | 0.00 | 0.00 |
| Good (G) | 0.75 | 0.10 | 0.15 |
| Medium Good (MG) | 0.60 | 0.30 | 0.10 |
| Medium (M) | 0.50 | 0.45 | 0.05 |
| Medium Poor (MP) | 0.30 | 0.50 | 0.20 |
| Poor (P) | 0.15 | 0.70 | 0.15 |
| Very Poor (VP) | 0.00 | 0.90 | 0.10 |

$$\text{IFWA}\left(\tilde{E}_1, \tilde{E}_2, \ldots, \tilde{E}_K\right) = \left(1 - \prod_{k=1}^{K}\left(1 - \mu_{ij}^{(k)}\right)^{\lambda_K}, \prod_{k=1}^{K}\left(v_{ij}^{(k)}\right)^{\lambda_K}\right). \tag{5.15}$$

**Step 5**: The decision matrix is normalized in this phase. This phase is performed following the procedure of [10].

**Step 6**: This step computes the weights of Fine–Kinney risk parameters using the IFWA operator as given in Step 4. The formulations for this computation are adapted from [3] and provided in Eqs. (5.16–5.18).

$$\tilde{w}_j = \left(1 - \prod_{k=1}^{K}\left(1 - \mu_{ij}^{(k)}\right)^{\lambda_K}, \prod_{k=1}^{K}\left(v_{ij}^{(k)}\right)^{\lambda_K}\right) \tag{5.16}$$

$$\tilde{w}_j = -\frac{1}{n \ln 2}\left(\mu_j \ln \mu_j + v_j \ln v_j - \left(1 - \pi_j\right)\ln\left(1 - \pi_j\right) - \pi_j \ln 2\right) \tag{5.17}$$

$$w_j = \frac{1 - \tilde{w}_j}{n - \sum_{j=1}^{n} \tilde{w}_j} \tag{5.18}$$

where $\sum_{j=1}^{n} w_j = 1$.

**Step 7**: The relative weights are computed as in Eq. (5.10) of the classical TODIM.

**Step 8**: The dominance of each hazard is established using Eqs. (5.19–5.20).

$$\Phi_j(H_i, H_l) = \begin{cases} \sqrt{\frac{w_{jr}}{\sum_{j=1}^{n} w_{jr}}} d\left(\tilde{r}_{ij}, \tilde{r}_{lj}\right) & \text{if } \tilde{r}_{ij} > \tilde{r}_{lj} \\ 0 & \text{if } \tilde{r}_{ij} = \tilde{r}_{lj}. \\ -\frac{1}{\theta}\sqrt{\frac{\sum_{j=1}^{n} w_{jr}}{w_{jr}}} d\left(\tilde{r}_{ij}\tilde{r}_{lj}\right) & \text{if } \tilde{r}_{ij} < \tilde{r}_{lj} \end{cases} \tag{5.19}$$

$$d\left(\tilde{r}_{ij},\tilde{r}_{lj}\right) = \sqrt{\frac{1}{2}\left(\mu_{ij} - \mu_{lj}\right)^2 + \left(v_{ij} - v_{lj}\right)^2 + \left(\pi_{ij} - \pi_{lj}\right)^2}. \tag{5.20}$$

**Step 9**: The dominance matrix is established as follows by Eq. (5.21).

$$\Phi_j = \left[\Phi^j_{il}\right]_{mxm} = \begin{pmatrix} \Phi^j_{11} & \cdots & \Phi^j_{1m} \\ \vdots & \ddots & \vdots \\ \Phi^j_{m1} & \cdots & \Phi^j_{mm} \end{pmatrix} \tag{5.21}$$

**Step 10**: Global dominance degree is obtained by using Eq. (5.22).

$$\xi(H_i, H_l) = \sum_{j=1}^{n} \Phi(H_i, H_l) \tag{5.22}$$

**Step 11**: The dominance measurements are then normalized as in Eq. (5.23).

$$\xi(H_i, H_l) = \sum_{l=1}^{m} \xi(H_i, H_l) - \underbrace{\min}_{i \in M} \sum_{l=1}^{m} \xi(H_i, H_l) / \underbrace{\max}_{i \in M} \sum_{l=1}^{m} \xi(H_i, H_l) - \underbrace{\min}_{i \in M} \sum_{l=1}^{m} \xi(H_i, H_l) \tag{5.23}$$

**Step 12**: Hazards are prioritized in descending order of $\xi$ values. The proposed approach's bird's eye view is also demonstrated in Fig. 5.1.

## 5.3 Case Study

A real-world case study in the gun and rifle assembly line of a weapon factory is provided to demonstrate the approach applicability. The observed factory under study is stationed in Istanbul/Turkey. It is the first factory in manufacturing semiauto shotguns throughout the country. It manufactures civilian and defense–law enforcements. Inside the manufacturing environment, a number of hazards have emerged in the assembly line because of product variety and rapid growing factors. The OHS experts made an initial evaluation in the assembly line and identified 19 hazards. The list of hazards is given in Table 5.2. A group consisting of three OHS experts participated in the risk assessment.

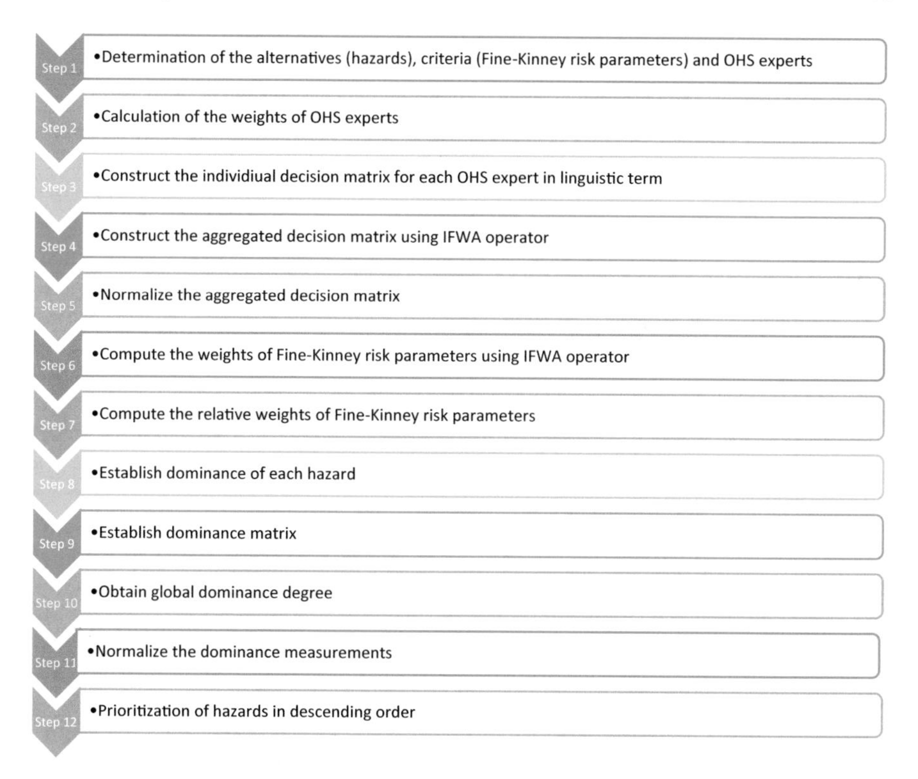


**Fig. 5.1** The proposed approach flowchart

### *5.3.1 Application Results*

The OHS risk assessment process for the observed assembly line is used by applying the given steps:

**Step 1**: Nineteen hazards are evaluated using three risk parameters of probability (P), exposure (E), and consequence (C) to assess the riskiness of the observed assembly line. Also, three OHS experts participate in this evaluation.

**Step 2**: Weights of OHS experts are calculated using Eq. (5.14). The results demonstrate that the weights of OHS experts are obtained as 0.254, 0.425, and 0.321. The highest weight is assigned to the second OHS expert and least important to the first OHS expert. The computations related to this step are given in Table 5.3.

**Step 3**: OHS ratings regarding the hazards according to three risk parameters are shown in Table 5.4. Using the scale and intuitionistic fuzzy numbers in Table 5.1, these linguistic judgments are transformed into intuitionistic fuzzy numbers.

**Step 4**: The aggregated decision matrix in intuitionistic fuzzy numbers is constructed using the IFWA operator. The results are given in Table 5.5.

**Table 5.2** Identified hazards with descriptions

| Code (H: Hazard) | Hazard identification | Possible effect | Current situation |
|---|---|---|---|
| H1 | Layout of the worksite (Tripping hazard) | Wound and bruise, injury | The shelves are not fixed |
| H2 | Slippery floor | Injuries | Pistol oil immersion is slippery due to running-in oil |
| H3 | Manual handling, lifting, placing, loading, and forcing | Joint, waist discomfort | Employees have not received training on *Manual Lifting and Moving* |
| H4 | Insufficient ventilation | Discomfort, stress, poisoning, and death | |
| H5 | Constant standing, constant sitting | Joint disorders, varicose veins, and stress | Staff were not trained on the ergonomics rules |
| H6 | Working with the hand tool | Cuts, injuries | Employees were not trained on the *Usage of Hand Tools* |
| H7 | Repeated movements | Injury, joint ailments, and stress | Staff were not trained on the ergonomics rules |
| H8 | Rotating—moving parts of the machine | Injury, death | Missing enclosures on some of the counters; damaged in some; and some also have swivels canceled. Employees do not have *Vocational Training*. Employees start working without being informed about the equipment |
| H9 | Electricity | Combustion, injury, and death | |
| H10 | Falling and flying objects | Injuries | Employees do not wear safety glasses during operations |
| H11 | Fire | Burn, drowning, and death | Emergency plan and teams are available, but current and education sensitivities are not taken into consideration |
| H12 | Unsuitable climatic conditions | Disease, injury | Especially in summer, it becomes a serious hot environment |

(continued)

**Table 5.2** (continued)

| Code (H: Hazard) | Hazard identification | Possible effect | Current situation |
|---|---|---|---|
| H13 | Noise | Hearing loss, stress/noise pollution | Noise measurements were made at the workplace. The use of headphones is good but not sufficient |
| H14 | Pointed/sharp area | Injury, death | |
| H15 | Brightness, excessive lighting, and insufficient lighting | Eye conditions, stress | There is sufficient lighting. However, it should be revised according to the lighting measurements to be made. Some lighting and mobile cables are not compatible with the standard |
| H16 | Energy cut-out (electric, pneumatic, hydraulic) | Burning, distortion, and death | |
| H17 | Exposure to chemical liquid and gas | Cancer, burns, eye conditions, and irritation | |
| H18 | Emergency events | Injury, death | Emergency plan and teams are available, but current and education sensitivities are not taken into consideration |
| H19 | Professional competence and experience | Injury, death | Employees have no Vocational Training |

**Table 5.3** Weight determination of OHS experts

| OHS expert | Linguistic term | Weight of expert ($\lambda_k$) |
|---|---|---|
| OHS expert-1 | M | 0.254 |
| OHS expert-2 | G | 0.425 |
| OHS expert-3 | MG | 0.321 |

**Step 5**: The matrix of decision is normalized in this step. It is executed following the procedure of [10]. It should be noted that all of the three risk parameters are benefit criteria.

**Step 6**: By using the expert judgments in Table 5.4 regarding Fine–Kinney risk parameters, this step computes the weights using the IFWA operator. The final weights of these parameters are given in Fig. 5.2. Highest importance is given to the consequence parameter with a value of 0.361, and the least importance is given to the probability parameter with a value of 0.300.

**Table 5.4** Ratings of hazards by OHS experts

| OHS expert-1 | | H1 | H2 | H3 | H4 | H5 | H6 | H7 | H8 | H9 | H10 | H11 | H12 | H13 | H14 | H15 | H16 | H17 | H18 | H19 |
|---|---|---|---|---|---|---|---|---|---|---|---|---|---|---|---|---|---|---|---|---|
| P | MP | MG | MG | MG | MG | M | MG | MG | M | MG | M | M | MG | M | MG | MG | M | MG | M | MG |
| E | G | G | G | G | G | G | G | VG | MG | P | P | VG | MG | G | MG | P | VP | MG | M | P |
| C | G | MG | MG | MG | G | MG | MG | MG | VG | VG | VG | G | MP | G | MG | MG | VG | G | G | G |
| OHS expert-2 | | H1 | H2 | H3 | H4 | H5 | H6 | H7 | H8 | H9 | H10 | H11 | H12 | H13 | H14 | H15 | H16 | H17 | H18 | H19 |
| P | M | MG | MG | MG | M | M | MG | MG | M | M | MG | M | MG | M | MG | M | MG | M | MG | M |
| E | MG | G | G | G | G | G | G | VG | MG | P | P | VG | MG | G | M | P | VP | MG | MG | P |
| C | G | MG | MG | MG | G | MG | MG | MG | VG | VG | VG | G | M | G | MG | MG | VG | G | G | G |
| OHS expert-3 | | H1 | H2 | H3 | H4 | H5 | H6 | H7 | H8 | H9 | H10 | H11 | H12 | H13 | H14 | H15 | H16 | H17 | H18 | H19 |
| P | M | MG | MG | MG | MG | M | MG | MG | M | MG | M | M | MG | M | MG | M | M | MG | M | M |
| E | G | G | G | G | G | G | G | VG | M | P | P | VG | MG | G | M | P | VP | MG | MG | P |
| C | G | MG | MG | MG | G | MG | MG | MG | VG | VG | VG | G | M | G | MG | MG | VG | G | G | G |

**Table 5.5** The aggregated intuitionistic fuzzy decision matrix

| Hazard | Probability | | | Exposure | | | Consequence | | |
|---|---|---|---|---|---|---|---|---|---|
| H1 | 0.600 | 0.300 | 0.100 | 0.750 | 0.100 | 0.150 | 0.600 | 0.300 | 0.100 |
| H2 | 0.600 | 0.300 | 0.100 | 0.750 | 0.100 | 0.150 | 0.600 | 0.300 | 0.100 |
| H3 | 0.600 | 0.300 | 0.100 | 0.750 | 0.100 | 0.150 | 0.600 | 0.300 | 0.100 |
| H4 | 0.600 | 0.300 | 0.100 | 0.750 | 0.100 | 0.150 | 0.750 | 0.100 | 0.150 |
| H5 | 0.500 | 0.450 | 0.050 | 0.750 | 0.100 | 0.150 | 0.600 | 0.300 | 0.100 |
| H6 | 0.600 | 0.300 | 0.100 | 0.750 | 0.100 | 0.150 | 0.600 | 0.300 | 0.100 |
| H7 | 0.600 | 0.300 | 0.100 | 1.000 | 0.000 | 0.000 | 0.600 | 0.300 | 0.100 |
| H8 | 0.500 | 0.450 | 0.050 | 0.570 | 0.352 | 0.084 | 1.000 | 0.000 | 0.000 |
| H9 | 0.560 | 0.368 | 0.079 | 0.150 | 0.700 | 0.150 | 1.000 | 0.000 | 0.000 |
| H10 | 0.545 | 0.391 | 0.072 | 0.150 | 0.700 | 0.150 | 1.000 | 0.000 | 0.000 |
| H11 | 0.500 | 0.450 | 0.050 | 1.000 | 0.000 | 0.000 | 0.750 | 0.100 | 0.150 |
| H12 | 0.600 | 0.300 | 0.100 | 0.600 | 0.300 | 0.100 | 0.455 | 0.463 | 0.091 |
| H13 | 0.500 | 0.450 | 0.050 | 0.750 | 0.100 | 0.150 | 0.750 | 0.100 | 0.150 |
| H14 | 0.600 | 0.300 | 0.100 | 0.528 | 0.415 | 0.063 | 0.600 | 0.300 | 0.100 |
| H15 | 0.528 | 0.415 | 0.063 | 0.150 | 0.700 | 0.150 | 0.600 | 0.300 | 0.100 |
| H16 | 0.545 | 0.391 | 0.072 | 0.000 | 0.900 | 0.100 | 1.000 | 0.000 | 0.000 |
| H17 | 0.560 | 0.368 | 0.079 | 0.600 | 0.300 | 0.100 | 0.750 | 0.100 | 0.150 |
| H18 | 0.545 | 0.391 | 0.072 | 0.577 | 0.342 | 0.088 | 0.750 | 0.100 | 0.150 |
| H19 | 0.528 | 0.415 | 0.063 | 0.150 | 0.700 | 0.150 | 0.750 | 0.100 | 0.150 |

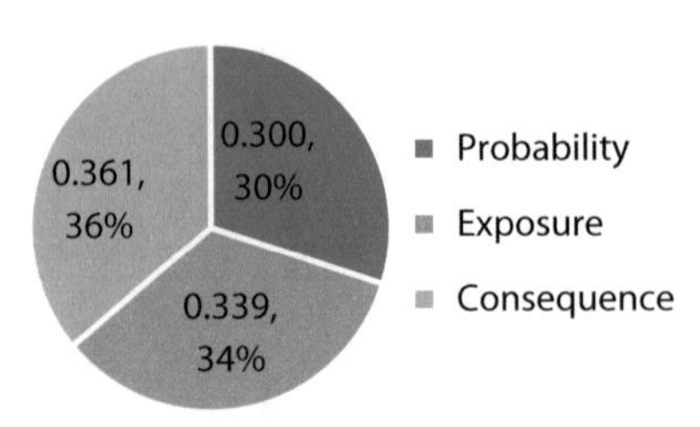


**Fig. 5.2** Weight values of Fine–Kinney risk parameters using IFWA operator

**Step 7**: The relative weights are obtained as 0.831, 0.940, and 1.000 for probability, exposure, and consequence, respectively, using Eq. (5.10).

**Step 8** and **Step 9**: The dominance of each hazard and dominance matrix is generated using the formulas as explained in the previous subsection.

**Step 10** and **Step 11**: The global dominance degrees and their normalized values are obtained as in Table 5.6.

**Step 12**: Hazards are prioritized. Figure 5.3 displays the rankings of the hazards.

**Table 5.6** The global dominance degrees and their normalized values

| Hazard | Global dominance degree | Normalized dominance measurement |
|---|---|---|
| H1 | 0.815 | 0.796 |
| H2 | 0.815 | 0.796 |
| H3 | 0.815 | 0.796 |
| H4 | 5.717 | 1.000 |
| H5 | −3.674 | 0.610 |
| H6 | 0.815 | 0.796 |
| H7 | 4.254 | 0.939 |
| H8 | 0.176 | 0.770 |
| H9 | −7.855 | 0.436 |
| H10 | −8.592 | 0.406 |
| H11 | 4.667 | 0.956 |
| H12 | −7.929 | 0.433 |
| H13 | 1.228 | 0.814 |
| H14 | −5.777 | 0.523 |
| H15 | −18.367 | 0.000 |
| H16 | −14.241 | 0.171 |
| H17 | 0.130 | 0.768 |
| H18 | −1.520 | 0.700 |
| H19 | −13.465 | 0.204 |

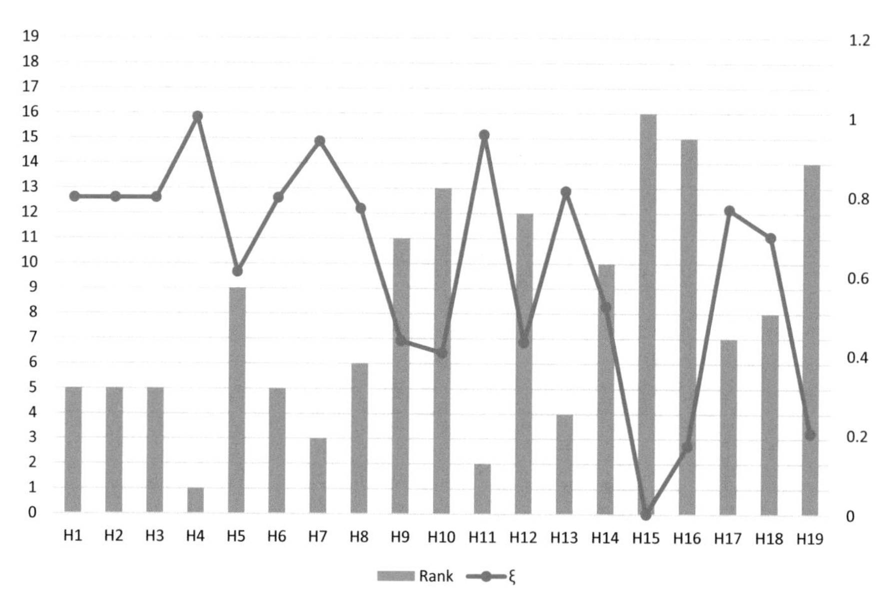


**Fig. 5.3** Final normalized dominance measurements and ranking orders of hazards

Results show that the most serious risks are stemmed from insufficient ventilation *H4*, fire *H11*, and repeated movements *H7*.

## 5.3.2 Validation Study on the Results

In this chapter, we made one sensitivity analysis to test the validity of the approach. The attenuation parameter "$\theta$" has been changed from 1 to 5. The normalized global performances and ranking of hazards for sensitivity analysis are shown in Table 5.7. It is clearly observed that the 19 hazard rankings for the observed assembly line are consistent. The ranking order of hazards has changed partially by varying the value of $\theta$ attenuation parameter.

**Table 5.7** The sensitivity analysis using the attenuation parameter

| Hazard | $\theta = 1$ | | $\theta = 2$ | | $\theta = 3$ | | $\theta = 4$ | | $\theta = 5$ | |
|---|---|---|---|---|---|---|---|---|---|---|
| | $\xi$ | Rank | $\xi$ | Rank | $\xi$ | Rank | $\xi$ | Rank | $\xi$ | Rank |
| H1 | 0.796 | 5 | 0.773 | 6 | 0.747 | 6 | 0.726 | 6 | 0.711 | 6 |
| H2 | 0.796 | 5 | 0.773 | 6 | 0.747 | 6 | 0.726 | 6 | 0.711 | 6 |
| H3 | 0.796 | 5 | 0.773 | 6 | 0.747 | 6 | 0.726 | 6 | 0.711 | 6 |
| H4 | 1.000 | 1 | 1.000 | 1 | 0.984 | 2 | 0.969 | 3 | 0.959 | 3 |
| H5 | 0.610 | 9 | 0.569 | 10 | 0.535 | 10 | 0.510 | 11 | 0.492 | 11 |
| H6 | 0.796 | 5 | 0.773 | 7 | 0.747 | 6 | 0.726 | 6 | 0.711 | 6 |
| H7 | 0.939 | 3 | 0.970 | 3 | 0.975 | 3 | 0.973 | 2 | 0.971 | 2 |
| H8 | 0.770 | 6 | 0.793 | 5 | 0.795 | 4 | 0.793 | 4 | 0.791 | 4 |
| H9 | 0.436 | 11 | 0.512 | 11 | 0.551 | 9 | 0.574 | 9 | 0.591 | 9 |
| H10 | 0.406 | 13 | 0.476 | 13 | 0.512 | 11 | 0.534 | 10 | 0.550 | 10 |
| H11 | 0.956 | 2 | 0.993 | 2 | 1.000 | 1 | 1.000 | 1 | 1.000 | 1 |
| H12 | 0.433 | 12 | 0.444 | 14 | 0.443 | 12 | 0.441 | 13 | 0.439 | 13 |
| H13 | 0.814 | 4 | 0.796 | 4 | 0.772 | 5 | 0.753 | 5 | 0.739 | 5 |
| H14 | 0.523 | 10 | 0.501 | 12 | 0.479 | 13 | 0.463 | 12 | 0.451 | 12 |
| H15 | 0.000 | 16 | 0.000 | 17 | 0.000 | 16 | 0.000 | 16 | 0.000 | 16 |
| H16 | 0.171 | 15 | 0.287 | 15 | 0.354 | 14 | 0.397 | 14 | 0.428 | 14 |
| H17 | 0.768 | 7 | 0.753 | 8 | 0.733 | 7 | 0.715 | 7 | 0.703 | 7 |
| H18 | 0.700 | 8 | 0.679 | 9 | 0.656 | 8 | 0.638 | 8 | 0.624 | 8 |
| H19 | 0.204 | 14 | 0.227 | 16 | 0.237 | 15 | 0.244 | 15 | 0.248 | 15 |

## 5.4 Python Implementation of the Proposed Approach

```
# Chapter 5
# import required libraries
import numpy as np

n_criteria = 3  # P, E, C
ns_value = 3  # μ, ν, π
n_expert = 3  # DM-1,DM-2, DM-3
n_hazard = 19
ln_val = -1 / (n_criteria * np.log(2))

linguistic_terms = [  # Linguistic term, μ, ν, π
    ["VG", 1.00, 0.00, 0.00],
    ["G", 0.75, 0.10, 0.15],
    ["MG", 0.60, 0.30, 0.10],
    ["M", 0.50, 0.45, 0.05],
    ["MP", 0.30, 0.50, 0.20],
    ["P", 0.15, 0.70, 0.15],
    ["VP", 0.00, 0.90, 0.10]
]

e_weight = [
    # experts_weight
    ["DM-1", "M", 0, 0],
    ["DM-2", "G", 0, 0],
    ["DM-3", "MG", 0, 0]
]

# elm: expert linguistic matrix
elm = [
    # id      P      E   C
    ["DM1", "MP", "G", "G"],
    ["DM2", "M", "MG", "G"],
    ["DM3", "M", "G", "G"]
]
```

```
# ehe: 1st expert's hazard evaluation
ehe1 = [
   # Hazard id,   P, E, C,
   ["H1", "MG", "G", "MG"],
   ["H2", "MG", "G", "MG"],
   ["H3", "MG", "G", "MG"],
   ["H4", "MG", "G", "G"],
   ["H5", "M", "G", "MG"],
   ["H6", "MG", "G", "MG"],
   ["H7", "MG", "VG", "MG"],
   ["H8ü", "M", "MG", "VG"],
   ["H9", "MG", "P", "VG"],
   ["H10", "M", "P", "VG"],
   ["H11", "M", "VG", "G"],
   ["H12", "MG", "MG", "MP"],
   ["H13", "M", "G", "G"],
   ["H14", "MG", "MG", "MG"],
   ["H15", "MG", "P", "MG"],
   ["H16", "M", "VP", "VG"],
   ["H17", "MG", "MG", "G"],
   ["H18", "M", "M", "G"],
   ["H19", "MG", "P", "G"]
 ]
 # ehe: 2nd expert's hazard evaluation
 ehe2 = [
   # Hazard id,P, E, C
   ["H1", "MG", "G", "MG"],
   ["H2", "MG", "G", "MG"],
   ["H3", "MG", "G", "MG"],
   ["H4", "MG", "G", "G"],
   ["H5", "M", "G", "MG"],
   ["H6", "MG", "G", "MG"],
   ["H7", "MG", "VG", "MG"],
   ["H8", "M", "MG", "VG"],
   ["H9", "M", "P", "VG"],
   ["H10", "MG", "P", "VG"],
   ["H11", "M", "VG", "G"],
   ["H12", "MG", "MG", "M"],
   ["H13", "M", "G", "G"],
   ["H14", "MG", "M", "MG"],
   ["H15", "M", "P", "MG"],
   ["H16", "MG", "VP", "VG"],
   ["H17", "M", "MG", "G"],
   ["H18", "MG", "MG", "G"],
   ["H19", "M", "P", "G"]
 ]
```

```
# ehe: 3rd expert's hazard evaluation
ehe3 = [
  # Hazard id, P, E, C
  ["H1", "MG", "G", "MG"],
  ["H2", "MG", "G", "MG"],
  ["H3", "MG", "G", "MG"],
  ["H4", "MG", "G", "G"],
  ["H5", "M", "G", "MG"],
  ["H6", "MG", "G", "MG"],
  ["H7", "MG", "VG", "MG"],
  ["H8", "M", "M", "VG"],
  ["H9", "MG", "P", "VG"],
  ["H10", "M", "P", "VG"],
  ["H11", "M", "VG", "G"],
  ["H12", "MG", "MG", "M"],
  ["H13", "M", "G", "G"],
  ["H14", "MG", "M", "MG"],
  ["H15", "M", "P", "MG"],
  ["H16", "M", "VP", "VG"],
  ["H17", "MG", "MG", "G"],
  ["H18", "M", "MG", "G"],
  ["H19", "M", "P", "G"]
]
def rank(decision_vector, direction): # direction -1:descending, 1:ascending
  order = np.zeros([len(decision_vector), 1])
  unique_val = direction * np.sort(direction * np.unique(decision_vector))
  for ix in range(0, len(unique_val)):
    order[np.argwhere(decision_vector == unique_val[ix])] = ix + 1
  return order


def print_result(order, vector):
  print('Hazard Id, Rank, Value')
  for ix in range(0, len(order)):
    print(ehe1[ix][0], ', ', int(order[ix]), ', ', vector[ix])

sm = 0
for j in range(0, n_expert):
  for fs in linguistic_terms:
    if e_weight[j][1] == fs[0]:
      e_weight[j][2] = fs[1] + fs[3] * (fs[1] / (1 - fs[3]))
      sm += e_weight[j][2]

for j in range(0, n_expert):
  e_weight[j][3] = e_weight[j][2] / sm # DM_weight
```

```
# e_numeric: numeric values of experts linguistic term
e_numeric = np.zeros([n_expert, n_criteria, ns_value], dtype=float)
for i in range(0, n_expert):
  for j in range(0, n_criteria):
    for lt in linguistic_terms:
      if elm[i][j + 1] == lt[0]:
        e_numeric[i][j] = lt[1:]

criteria_matrix = np.zeros([n_criteria, ns_value], dtype=float)
for j in range(0, n_criteria):

for k in range(0, ns_value):
  val = 1
  for i in range(0, n_expert):
    if k == 0:
      val *= np.power(1 - e_numeric[i][j][k], e_weight[i][3])
    elif k == 1:
      val *= np.power(e_numeric[i][j][k], e_weight[i][3])
    elif k == 2:
      val = 1 - criteria_matrix[j][0] - criteria_matrix[j][1]
  if k == 0:
    val = 1 - val
  criteria_matrix[j][k] = val
  criteria_matrix2 = np.zeros([n_criteria, ns_value + 1], dtype=float)
  for j in range(0, n_criteria):
    for k in range(0, ns_value):
      if k == 0 or k == 1:
        criteria_matrix2[j][k] = criteria_matrix[j][k] * np.log(criteria_matrix[j][k])
      elif k == 2:
        val2 = criteria_matrix[j][k] * np.log(2)
        criteria_matrix2[j][k + 1] = val2
        val = 1 - criteria_matrix[j][k]
        criteria_matrix2[j][k] = val * np.log(val)
        criteria_matrix3 = []
        for cm in criteria_matrix2:
          criteria_matrix3.append(ln_val * (cm[0] + cm[1] - cm[2] - cm[3]))
```

```
criteria_matrix4 = []
sm = np.sum(criteria_matrix3)
for cm in criteria_matrix3:
    criteria_matrix4.append((1 - cm) / (3 - sm))

relative_weights = []
mx = np.max(criteria_matrix4)
for cm in criteria_matrix4:
    relative_weights.append(cm / mx)

ehe = np.asarray([ehe1, ehe2, ehe3])
ehe_numeric = np.zeros([n_expert, n_hazard, n_criteria, ns_value], dtype=float)
for i in range(0, n_expert):
    for j in range(0, n_hazard):
        for k in range(0, n_criteria):
                    for lt in linguistic_terms:
                        if ehe[i][j][k + 1] == lt[0]:
                            ehe_numeric[i][j][k] = lt[1:]

aggregated_matrix = np.ones([n_hazard, n_criteria, ns_value], dtype=float)
for i in range(0, n_expert):
    aggregated_matrix *= np.power(1 - ehe_numeric[i], e_weight[i][3])
aggregated_matrix = 1 - aggregated_matrix

theta = 1
mx = np.max(aggregated_matrix, axis=0)[0][0]
sm_weight = np.sum(relative_weights)
pairing_matrix = np.zeros([n_hazard, n_hazard, n_criteria, 2], dtype=float)
last_pair = np.zeros([n_hazard, n_hazard], dtype=float)
for i in range(0, n_hazard):
    for j in range(0, n_hazard):
        for k in range(0, n_criteria):
            val = 0
        for l in range(0, ns_value):
            val += np.square(aggregated_matrix[i][k][l] - aggregated_matrix[j][k][l])
        pairing_matrix[i][j][k][0] = np.sqrt(0.5 * val)
        if aggregated_matrix[i][k][0] > aggregated_matrix[j][k][0]:
            weight_1 = relative_weights[k] / sm_weight
            pairing_matrix[i][j][k][1] = np.sqrt(weight_1 * pairing_matrix[i][j][k][0])
        elif aggregated_matrix[i][k][0] < aggregated_matrix[j][k][0]:
                        weight_2 = sm_weight / relative_weights[k]
                        pairing_matrix[i][j][k][1] = np.sqrt(weight_2) * (-1 / theta) * pairing_matrix[i][j][k][0]
                    else:
                        pairing_matrix[i][j][k][1] = 0
                last_pair[i][j] = np.sum(pairing_matrix[i, j, :, 1])
```

```
sum_pair = np.sum(last_pair, axis=1)
mx = np.max(sum_pair)
mn = np.min(sum_pair)
# nmd: normalized measurement dominance
nmd = (sum_pair - mn) / (mx - mn)
hazard_rank = rank(nmd, -1)
print_result(hazard_rank, nmd)
'''
Output:
Hazard Id, Rank, Value
H1 , 5 , 0.7964755257526909
H2 , 5 , 0.7964755257526909
H3 , 5 , 0.7964755257526909
H4 , 1 , 1.0
H5 , 9 , 0.6100923168434228
H6 , 5 , 0.7964755257526909
H7 , 3 , 0.9392790956145556
H8ü , 6 , 0.76991876112814
H9 , 11 , 0.4364727266641372
H10 , 13 , 0.4058507512060441
H11 , 2 , 0.9564203609525965
H12 , 12 , 0.4333906372458517
H13 , 4 , 0.8136167910907319
H14 , 10 , 0.5227412425395443
H15 , 16 , 0.0
H16 , 15 , 0.17129408867692442
H17 , 7 , 0.7680156730371516
H18 , 8 , 0.699515430751146
H19 , 14 , 0.20352447424730905
'''
```

## References


1. Atanassov, K. T. (1999). Intuitionistic fuzzy sets. In *Intuitionistic fuzzy sets* (pp. 1–137). Heidelberg: Physica.
2. Xu, Z., & Liao, H. (2013). Intuitionistic fuzzy analytic hierarchy process. *IEEE Transactions on Fuzzy Systems, 22*(4), 749–761.
3. Abdullah, L., & Najib, L. (2014). A new preference scale of intuitionistic fuzzy analytic hierarchy process in multi-criteria decision making problems. *Journal of Intelligent & Fuzzy Systems, 26*(2), 1039–1049.
4. Liao, H., Mi, X., Xu, Z., Xu, J., & Herrera, F. (2018). Intuitionistic fuzzy analytic network process. *IEEE Transactions on Fuzzy Systems, 26*(5), 2578–2590.
5. Boran, F. E., Genç, S., Kurt, M., & Akay, D. (2009). A multi-criteria intuitionistic fuzzy group decision making for supplier selection with TOPSIS method. *Expert Systems with Applications, 36*(8), 11363–11368.
6. Devi, K. (2011). Extension of VIKOR method in intuitionistic fuzzy environment for robot selection. *Expert Systems with Applications, 38*(11), 14163–14168.

7. Liao, H., & Xu, Z. (2014). Multi-criteria decision making with intuitionistic fuzzy PROMETHEE. *Journal of Intelligent & Fuzzy Systems, 27*(4), 1703–1717.
8. Wu, M. C., & Chen, T. Y. (2011). The ELECTRE multicriteria analysis approach based on Atanassov's intuitionistic fuzzy sets. *Expert Systems with Applications, 38*(10), 12318–12327.
9. Govindan, K., Khodaverdi, R., & Vafadarnikjoo, A. (2015). Intuitionistic fuzzy based DEMATEL method for developing green practices and performances in a green supply chain. *Expert Systems with Applications, 42*(20), 7207–7220.
10. Göçer, F., & Büyüközkan, G. (2019, July). Assessment of big data vendors by intuitionistic fuzzy TODIM. In *International Conference on Intelligent and Fuzzy Systems* (pp. 574–582). Cham: Springer.
11. Qin, Q., Liang, F., Li, L., Chen, Y. W., & Yu, G. F. (2017). A TODIM-based multi-criteria group decision making with triangular intuitionistic fuzzy numbers. *Applied Soft Computing, 55,* 93–107.
12. Krohling, R. A., Pacheco, A. G., & Siviero, A. L. (2013). IF-TODIM: An intuitionistic fuzzy TODIM to multi-criteria decision making. *Knowledge-Based Systems, 53,* 142–146.
13. Gomes, L. F. A. M., & Lima, M. M. P. P. (1992). TODIM: Basics and application to multicriteria ranking of projects with environmental impacts. *Foundations of Computing and Decision Sciences, 16*(4), 113–127.
14. Huang, J., Li, Z. S., & Liu, H. C. (2017). New approach for failure mode and effect analysis using linguistic distribution assessments and TODIM method. *Reliability Engineering & System Safety, 167,* 302–309.
15. Soni, N., Christian, R. A., & Jariwala, N. (2016). Pollution potential ranking of industries using classical TODIM method. *Journal of Environmental Protection, 7*(11), 1645.
16. Ozdemir, Y., & Gul, M. (2019). Measuring development levels of NUTS-2 regions in Turkey based on capabilities approach and multi-criteria decision-making. *Computers & Industrial Engineering, 128,* 150–169.

# Chapter 6
# Fine–Kinney-Based Occupational Risk Assessment Using Hexagonal Fuzzy MULTIMOORA

**Abstract** Hexagonal fuzzy numbers (HFNs) can be used as a proficient logic to simplify understanding of ambiguity information. HFNs present the usual information in a comprehensive way and also the ambiguity section can be exemplified in a reasonable way. In this chapter, we proposed an improved Fine–Kinney occupational risk assessment approach using a well-known MCDM method Multi-Objective Optimization by Ratio Analysis (MULTIMOORA) using hexagonal fuzzy numbers. Since the mere MULTIMOORA has failed to handle uncertainty and vague information which usually exist in real world problems, we follow integration of HFNs and MULTIMOORA (HFMULTIMOORA). To show the applicability of the novel approach, a case study of risk assessment of a raw mill in cement plant was provided. Comparative analysis with using two aggregation tools as reciprocal rank method and dominance theory are carried out. Finally, the Python implementation of the proposed approach is implemented to be effective for those concerned in the future.

## 6.1 Hexagonal Fuzzy Numbers and MULTIMOORA

### *6.1.1 Hexagonal Fuzzy Numbers and Its Operations*

Hexagonal Fuzzy Number (HFN) are considered and applied on decision making problem by Rajarajeswari et al. [1], Dinagar and Narayanan [2] and Rajarajeswari and Sudha [3]. HFN is specified by six numbers [4] as $\widetilde{A} = (a_1, a_2, a_3, a_4, a_5, a_6)$ where $a_1, a_2, a_3, a_4, a_5$, and $a_6$ are crisp numbers and $a_1 \leq a_2 \leq a_3 \leq a_4 \leq a_5 \leq a_6$ with membership function is presented below:



M. Gul et al., *Fine–Kinney-Based Fuzzy Multi-criteria Occupational Risk Assessment*, Studies in Fuzziness and Soft Computing 398,
https://doi.org/10.1007/978-3-030-52148-6_6

$$\mu_{\tilde{A}}(x) = \begin{cases} \frac{1}{2}\left(\frac{x - a_1}{a_2 - a_1}\right), & a_1 \le x \le a_2 \\ \frac{1}{2} + \frac{1}{2}\left(\frac{x - a_2}{a_3 - a_2}\right), & a_2 \le x \le a_3 \\ 1, & a_3 \le x \le a_4 \\ 1 - \frac{1}{2}\left(\frac{x - a_4}{a_5 - a_4}\right), & a_4 \le x \le a_5 \\ \frac{1}{2}\left(\frac{a_6 - x}{a_6 - a_5}\right), & a_5 \le x \le a_6 \\ 0, & otherwise \end{cases} \tag{6.1}$$

$\mu_{\tilde{A}}(x)$ is a continuous function in [0, 1]. If two positive HFN $\widetilde{A} = (a_1, a_2, a_3, a_4, a_5, a_6)$ and $\widetilde{B} = (b_1, b_2, b_3, b_4, b_5, b_6)$ are considered, operations on these two positive HFNs are calculated [5, 6] as follows:

Addition:

$$\widetilde{A} + \widetilde{B} = (a_1 + b_1, a_2 + b_2, a_3 + b_3, a_4 + b_4, a_5 + b_5, a_6 + b_6) \tag{6.2}$$

Subtraction:

$$\widetilde{A} - \widetilde{B} = (a_1 - b_1, a_2 - b_2, a_3 - b_3, a_4 - b_4, a_5 - b_5, a_6 - b_6) \tag{6.3}$$

Multiplication:

$$\widetilde{A} \times \widetilde{B} = (a_1 \times b_1, a_2 \times b_2, a_3 \times b_3, a_4 \times b_4, a_5 \times b_5, a_6 \times b_6) \tag{6.4}$$

Division:

$$\widetilde{A}/\widetilde{B} = (a_1/b_6, a_2/b_5, a_3/b_4, a_4/b_3, a_5/b_2, a_6/b_1) \tag{6.5}$$

If $\widetilde{A} = (a_1, a_2, a_3, a_4, a_5, a_6)$ and $\widetilde{B} = (b_1, b_2, b_3, b_4, b_5, b_6)$ are two HFNs, then the hamming distance of $\widetilde{A}$ from $\widetilde{B}$ is given,

$$d\left(\widetilde{A}, \widetilde{B}\right) = \frac{1}{6}(|a_1 - b_1| + |a_2 - b_2| + |a_3 - b_3| + |a_4 - b_4| + |a_5 - b_5| + |a_6 - b_6|) \tag{6.6}$$

### 6.1.2 MULTIMOORA Method

The procedural steps of the base MULTIMOORA which was initially proposed by Brauers and Zavadskas [7] are summarized as follows [8]:

1. Construct a decision matrix and weight vector that is composed of the evaluation $m$ hazards of the problem with respect to $n$ parameters,
2. Normalize the decision matrix which is provided in Step 1,
3. Calculate the utility of Ratio System using the weighted normalized decision matrix,
4. Determine the most critical hazard based on Reference Point Approach
5. Determine the most critical hazard based on Full Multiplicative Form
6. Rank the alternatives using dominance theory, mathematical operators, MCDM approaches or programming approaches.

## 6.2 Proposed Fine–Kinney-Based Approach Using HFMULTIMOORA

The base MULTIMOORA method was integrated with fuzzy sets and its extensions such as triangular fuzzy sets [9–11], interval-valued fuzzy sets [12], neutrosophic sets [13], interval type-2 fuzzy sets [14, 15], hesitant fuzzy sets [16, 17], intuitionistic fuzzy sets [18] and pythagorean fuzzy sets [19, 20], interval-valued intuitionistic [21]. The detailed literature review of MULTIMOORA are presented by Baležentis and Baležentis [22] and Hafezalkotob et al. [8]. The interested readers and researchers can be analyzed this literature review. The detailed procedural flow of HFN MULTIMOORA is as follows:

**Step 1**: First step generates the decision matrix under hexagonal fuzzy number. It is denoted with $D_{\tilde{n}} = \left\langle d_{ij}^{s} \right\rangle_{mxn} = \left\langle p_{ij}, q_{ij}, r_{ij}, s_{ij}, t_{ij}, u_{ij} \right\rangle_{mxn}$ Its matrix notation is as in Eq. (6.7).

$$\begin{array}{c} \\ A_1 \\ A_2 \\ \ldots \\ A_m \end{array} \begin{array}{c} \begin{array}{cccc} C_1 & C_2 & \ldots & C_n \end{array} \\ \left( \begin{array}{cccc} (p_{11}, q_{11}, r_{11}, s_{11}, t_{11}, u_{11}) & (p_{12}, q_{12}, r_{12}, s_{12}, t_{12}, u_{12}) & \ldots & (p_{1n}, q_{1n}, r_{1n}, s_{1n}, t_{1n}, u_{1n}) \\ (p_{21}, q_{21}, r_{21}, s_{21}, t_{21}, u_{21}) & (p_{22}, q_{22}, r_{22}, s_{22}, t_{22}, u_{22}) & \ldots & (p_{2n}, q_{2n}, r_{2n}, s_{2n}, t_{2n}, u_{2n}) \\ \ldots & \ldots & \ldots & \ldots \\ (p_{m1}, q_{m1}, r_{m1}, s_{m1}, t_{m1}, u_{m1}) & (p_{m2}, q_{m2}, r_{m2}, s_{m2}, t_{m2}, u_{m2}) & \ldots & (p_{mn}, q_{mn}, r_{mn}, s_{mn}, t_{mn}, u_{mn}) \end{array} \right) \end{array} \quad (6.7)$$

In this step, the $d_{ij}^{s}$ presents the rating of the $i$th risk of the $j$th parameter evaluation by $s$th OHS expert where $i = 1, 2, \ldots, m; j = 1, 2, \ldots n; s = 1, 2, \ldots, S$

If the number of OHS experts more than one, then the normalization procedure is applied using Eq. (6.8) as follows:

$$\widetilde{d}_{ij} = \sum_{s=1}^{S} \widetilde{W}_s \widetilde{d}_{ij}^s / \sum_{s=1}^{S} \widetilde{W}_s \tag{6.8}$$

$\widetilde{W}_s$ the fuzzy coefficient of the $s$th OHS expert.

### 6.2.1 The Fuzzy Ratio System

The Ratio System specifies normalization of the HFNs $\widetilde{d}_{ij}$ emerging from matrix. The normalization is applied for HFN as follows:

$$\widetilde{d}_{ij}^* = \left(p_{ij}^*, q_{ij}^*, r_{ij}^*, s_{ij}^*, t_{ij}^*, u_{ij}^*\right)$$
$$= \begin{cases} p_{ij1}^* = p_{ij1} \Big/ \sqrt{\sum_{i=1}^{m} \left[\left(p_{ij1}^2\right) + \left(q_{ij1}^2\right) + \left(r_{ij1}^2\right) + \left(s_{ij1}^2\right) + \left(t_{ij1}^2\right) + \left(u_{ij1}^2\right)\right]} \\ q_{ij1}^* = q_{ij1} \Big/ \sqrt{\sum_{i=1}^{m} \left[\left(p_{ij1}^2\right) + \left(q_{ij1}^2\right) + \left(r_{ij1}^2\right) + \left(s_{ij1}^2\right) + \left(t_{ij1}^2\right) + \left(u_{ij1}^2\right)\right]} \\ r_{ij1}^* = r_{ij1} \Big/ \sqrt{\sum_{i=1}^{m} \left[\left(p_{ij1}^2\right) + \left(q_{ij1}^2\right) + \left(r_{ij1}^2\right) + \left(s_{ij1}^2\right) + \left(t_{ij1}^2\right) + \left(u_{ij1}^2\right)\right]} \\ s_{ij1}^* = s_{ij1} \Big/ \sqrt{\sum_{i=1}^{m} \left[\left(p_{ij1}^2\right) + \left(q_{ij1}^2\right) + \left(r_{ij1}^2\right) + \left(s_{ij1}^2\right) + \left(t_{ij1}^2\right) + \left(u_{ij1}^2\right)\right]} \\ t_{ij1}^* = t_{ij1} \Big/ \sqrt{\sum_{i=1}^{m} \left[\left(p_{ij1}^2\right) + \left(q_{ij1}^2\right) + \left(r_{ij1}^2\right) + \left(s_{ij1}^2\right) + \left(t_{ij1}^2\right) + \left(u_{ij1}^2\right)\right]} \\ u_{ij1}^* = u_{ij1} \Big/ \sqrt{\sum_{i=1}^{m} \left[\left(p_{ij1}^2\right) + \left(q_{ij1}^2\right) + \left(r_{ij1}^2\right) + \left(s_{ij1}^2\right) + \left(t_{ij1}^2\right) + \left(u_{ij1}^2\right)\right]} \end{cases} \forall i, j \tag{6.9}$$

Then, the weighted normalized decision matrix is calculated using the normalized matrix and the importance weighted of risk parameters.

$$\widetilde{e}_i^* = w_j * \widetilde{d}_{ij}^* \tag{6.10}$$

The summarizing ratios $\widetilde{e}_i^*$ is computed of for each $i$th risk. The summation and extraction operation are applied using Eqs. (6.2) or (6.3) respectively.

$$\widetilde{f}_i^* = \sum_{j=1}^{g} \widetilde{d}_{ij}^* - \sum_{j=g+1}^{n} \widetilde{d}_{ij}^* \tag{6.11}$$

where $g = 1, 2, ..., n$ symbolize the number of parameters to be maximized. Then, the defuzzfication procedure is applied using Dhurai and Karpagam [23] method for $\widetilde{e}_i^* = (\widetilde{f}_{i1}^*, \widetilde{f}_{i2}^*, \widetilde{f}_{i3}^*, \widetilde{f}_{i4}^*, \widetilde{f}_{i5}^*, \widetilde{f}_{i6}^*)$ as follows:

$$FRS_i = (\widetilde{f}_{i1}^* + 2\widetilde{f}_{i2}^* + \widetilde{f}_{i3}^* + \widetilde{f}_{i4}^* + 2\widetilde{f}_{i5}^* + f_{i6}^*)/4 \tag{6.12}$$

where $FRS_i$ symbolizes the best defuzzified value of the $i$th risk. Consequently, the risk with higher FRS values is ranked higher order.

### 6.2.2 The Fuzzy Reference Point

The fuzzy Reference Point approach is calculated using the fuzzy ratio system. The Maximal Objective Reference Point $\widetilde{M}_r$ is obtained with respect to Eq. (6.10). The $j$th risk parameter of the reference point is like the fuzzy maximum or minimum of the $j$th criterion $\widetilde{e}_i^+$, where

$$\begin{cases} \widetilde{e}_i^+ = \left(\max p_{ij}^*, \max q_{ij}^*, \max r_{ij}^*, \max s_{ij}^*, \max t_{ij}^*, \max u_{ij}^*\right), j \leq g; \\ \widetilde{e}_i^+ = \left(\min p_{ij}^*, \min q_{ij}^*, \min r_{ij}^*, \min s_{ij}^*, \min t_{ij}^*, \min u_{ij}^*\right) j > g. \end{cases}$$

Then, the distances between the maximal objective point and the weighted normalized matrix are calculated for all risk and parameters using Eq. (6.6). The Min–Max Metric of Tchebycheff is applied for obtaining the final ranks.

$$\min_i \left( \max_i \left| \widetilde{M}_r - \widetilde{e}_i^* \right| \right)$$

### 6.2.3 The Fuzzy Full Multiplicative Form

Overall utility of the $i$th risk can be stated as dimensionless number by using Eq. (6.5):

$$\widetilde{U}_i^* = \widetilde{A}_i / \widetilde{B}_i$$

where $\widetilde{A}_i = (A_{i1}, A_{i2}, A_{i3}, A_{i4}, A_{i5}, A_{i6}) = \prod_{j=1}^{g} d_{ij}, i = 1, 2, \ldots, m$ symbolizes element of parameters of the $i$th risk to be maximized with $g = 1, ..., n$ being the number of parameters to be maximized and where $\widetilde{B}_i = (B_{i1}, B_{i2}, B_{i3}, B_{i4}, B_{i5}, B_{i6}) = \prod_{j=1}^{g} d_{ij}, i = 1, 2, \ldots, m$ symbolizes element of parameters of the $i$th risk to be maximized with $n - g$ being the number of parameters to

**Implementing the HFMULTIMOORA**

- Generates the decision matrix under hexagonal fuzzy number
- Normalize the fuzzy numbers of the Ratio System
- Obtain the the weighted normalized decision matrix
- Calculate the Ratio System
- Calculate the fuzzy Reference Point approach
- Calculate the fuzzy Full Multiplicative Form
- Obtain the rank of HFMULTIMOORA using the reciprocal rank method

**Validation tests**

- Analyze results of hazard rankings in HFMULTIMOORA using the reciprocal rank method
- Obtain the HFMULTIMOORA using the dominance theory
- Test Spearman correlation analysis

**Fig. 6.1** The main steps of the proposed approach

be to be minimized. Then, the fuzzy utility is defuzzified and it is used to rank the risks.

At the final step of the fuzzy MULTIMOORA, we used reciprocal rank method. It considers the position of each alternative with respect to ranking of fuzzy ratio system $(r(y_i))$, fuzzy reference point $(r(z_i))$, and fuzzy full multiplicative form $(r(U_i))$. The score is calculated [24] as follows:

$$RPM\,(H_i) = 1/\left(\frac{1}{r(y_i)} + \frac{1}{r(z_i)} + \frac{1}{r(U_i)}\right)$$

The most critical risk is obtained based on rank position method which has the minimum value of reciprocal rank method.

The main steps of proposed approach are graphically demonstrated in Fig. 6.1. There are two main steps and their sub-steps through the study. The first is implementing MULTIMOORA under HFNs. Second is about the comparison of the proposed approach between classical Fine–Kinney and HFMULTIMOORA using dominance theory.

## 6.3 Case Study

To demonstrate the applicability of the proposed approach, a case study for occupational risk assessment of raw mill in cement plant is carried out. The implementation of proposed HFMULTIMOORA under Fine–Kinney approach is given in detail in the following sub-sections.

**Table 6.1** Hazard list in the observed raw mill in cement plant*

| Hazards | Activity area | Hazard identification | Associated risk definition |
|---|---|---|---|
| H1 | Reclaimer getting away from car oil and getting closer | Rotating-moving parts of machinery and parts | Injury, jamming, crushing, death |
| H2 | Cleaning of the front reclaimer close to the right part of the bucket | Rotating-moving parts of machinery and parts | Injury, jamming, crushing, death |
| H3 | | Slippery floor, insufficient and irregular work area | Running disruption, falling, injury and limb loss |
| H4 | | Non-ergonomic equipment/movement | Injury, joint and muscle disorders |
| H5 | Cleaning the reclaimer chute | Falling from high | Injury |
| H6 | | Non-ergonomic equipment/movement | Injury, joint and muscle disorders |
| H7 | | Slippery floor, insufficient and irregular work area | Running disruption, falling, injury and limb loss |
| H8 | | Manual handling, lifting, placing, loading, forcing | Joint and lumbar discomfort, compression and crushing |
| H9 | 2009 Cleaning of tape rollers and under tape | Rotating-moving parts of machinery and parts | Injury, jamming, crushing, death |
| H10 | | Slippery floor, insufficient and irregular work area | Falling and injury |
| H11 | | Non-ergonomic equipment/movement | Injury, joint and muscle disorders |
| H12 | | Falling from high | Injury |
| H13 | | Fixed and mobile ladder | Falling, stuttering and injury |
| H14 | | Unaware patrol | Injury, limb loss and death |
| H15 | 2010 Cleaning of tape rollers and under tape | Rotating-moving parts of machinery and parts | Injury, jamming, crushing, hand-arm foot grabber, limb loss |
| H16 | | Slippery floor, insufficient and irregular work area | Falling and injury |

(continued)

### *6.3.1 Application Results*

In this case study, twenty-six hazards and their associated risks are evaluated by the proposed approach with respect to the three Fine–Kinney parameters. The hazards which are indicted by "H" are adapted from [25]. The most important hazard sources and risks defined by OHS experts in the observed raw mill in cement plant are given in Table 6.1.

**Table 6.1** (continued)

| Hazards | Activity area | Hazard identification | Associated risk definition |
|---|---|---|---|
| H17 | | Non-ergonomic equipment/movement | Injury, joint and muscle disorders |
| H18 | | Falling from high | Injury |
| H19 | | Fixed and mobile ladder | Falling, stuttering and injury |
| H20 | | Unaware patrol | Injury, limb loss and death |
| H21 | 2010.1 Cleaning of tape rollers and under tape | Rotating-moving parts of machinery and parts | Injury, jamming, crushing, hand-arm foot grabber, limb loss |
| H22 | | Slippery floor, insufficient and irregular work area | Falling and injury |
| H23 | | Non-ergonomic equipment/movement | Injury, joint and muscle disorders |
| H24 | | Falling from high | Injury |
| H25 | | Unaware patrol | Injury and limb loss |
| H26 | | Fixed and mobile ladder | Falling, stuttering and injury |

**Table 6.2** Linguistic terms and their associated HF numbers

| Linguistic variables | Hexagonal fuzzy numbers |
|---|---|
| Very low (VL) | (1, 3, 5, 7, 9, 11) |
| Low (L) | (2, 4, 6, 8, 10, 12) |
| Medium low (ML) | (3, 5, 7, 9, 11, 13) |
| Medium high (MH) | (5, 7, 9, 11, 13, 15) |
| High (H) | (6, 8, 10, 12, 14, 16) |
| Very high (VH) | (7, 9, 11, 13, 15, 17) |

In the first of the proposed approach, the decision matrix for all hazards with respect to three parameters of Fine–Kinney are constructed. The decision matrix is constructed using Eq. (6.7). The linguistic variables and its HFN are used in this process that is presented in Table 6.2. The detailed decision matrix is given in Table 6.3.

Then, the normalization procedure is applied using Eq. (6.9), and it is presented in Table 6.4. Then, the weighted normalized decision matrix is calculated using the normalized matrix and the importance weighted of risk parameters. The importance weights of the three parameters are considered as 0.2890, 0.2930, and 0.4180 for probability, exposure, consequence, respectively. The weights of these parameters are derived from [26].

---

**Table 6.3** The hexagonal fuzzy decision matrix with respect to Fine–Kinney risk parameters

| | Probability | Exposure | Consequence |
|---|---|---|---|
| H1 | (1, 3, 5, 7, 9, 11) | (2, 4, 6, 8, 10, 12) | (6, 8, 10, 12, 14, 16) |
| H2 | (2, 4, 6, 8, 10, 12) | (6, 8, 10, 12, 14, 16) | (6, 8, 10, 12, 14, 16) |
| H3 | (2, 4, 6, 8, 10, 12) | (5, 7, 9, 11, 13, 15) | (6, 8, 10, 12, 14, 16) |
| H4 | (3, 5, 7, 9, 11, 13) | (6, 8, 10, 12, 14, 16) | (1, 3, 5, 7, 9, 11) |
| H5 | (3, 5, 7, 9, 11, 13) | (5, 7, 9, 11, 13, 15) | (3, 5, 7, 9, 11, 13) |
| H6 | (5, 7, 9, 11, 13, 15) | (6, 8, 10, 12, 14, 16) | (1, 3, 5, 7, 9, 11) |
| H7 | (2, 4, 6, 8, 10, 12) | (6, 8, 10, 12, 14, 16) | (3, 5, 7, 9, 11, 13) |
| H8 | (3, 5, 7, 9, 11, 13) | (6, 8, 10, 12, 14, 16) | (1, 3, 5, 7, 9, 11) |
| H9 | (2, 4, 6, 8, 10, 12) | (2, 4, 6, 8, 10, 12) | (6, 8, 10, 12, 14, 16) |
| H10 | (3, 5, 7, 9, 11, 13) | (6, 8, 10, 12, 14, 16) | (2, 4, 6, 8, 10, 12) |
| H11 | (5, 7, 9, 11, 13, 15) | (5, 7, 9, 11, 13, 15) | (2, 4, 6, 8, 10, 12) |
| H12 | (2, 4, 6, 8, 10, 12) | (5, 7, 9, 11, 13, 15) | (1, 3, 5, 7, 9, 11) |
| H13 | (3, 5, 7, 9, 11, 13) | (5, 7, 9, 11, 13, 15) | (2, 4, 6, 8, 10, 12) |
| H14 | (2, 4, 6, 8, 10, 12) | (6, 8, 10, 12, 14, 16) | (6, 8, 10, 12, 14, 16) |
| H15 | (2, 4, 6, 8, 10, 12) | (2, 4, 6, 8, 10, 12) | (6, 8, 10, 12, 14, 16) |
| H16 | (3, 5, 7, 9, 11, 13) | (6, 8, 10, 12, 14, 16) | (2, 4, 6, 8, 10, 12) |
| H17 | (5, 7, 9, 11, 13, 15) | (5, 7, 9, 11, 13, 15) | (2, 4, 6, 8, 10, 12) |
| H18 | (2, 4, 6, 8, 10, 12) | (5, 7, 9, 11, 13, 15) | (1, 3, 5, 7, 9, 11) |
| H19 | (3, 5, 7, 9, 11, 13) | (5, 7, 9, 11, 13, 15) | (2, 4, 6, 8, 10, 12) |
| H20 | (2, 4, 6, 8, 10, 12) | (6, 8, 10, 12, 14, 16) | (6, 8, 10, 12, 14, 16) |
| H21 | (2, 4, 6, 8, 10, 12) | (2, 4, 6, 8, 10, 12) | (6, 8, 10, 12, 14, 16) |
| H22 | (3, 5, 7, 9, 11, 13) | (6, 8, 10, 12, 14, 16) | (2, 4, 6, 8, 10, 12) |
| H23 | (5, 7, 9, 11, 13, 15) | (5, 7, 9, 11, 13, 15) | (2, 4, 6, 8, 10, 12) |
| H24 | (2, 4, 6, 8, 10, 12) | (5, 7, 9, 11, 13, 15) | (1, 3, 5, 7, 9, 11) |
| H25 | (2, 4, 6, 8, 10, 12) | (6, 8, 10, 12, 14, 16) | (6, 8, 10, 12, 14, 16) |
| H26 | (3, 5, 7, 9, 11, 13) | (5, 7, 9, 11, 13, 15) | (2, 4, 6, 8, 10, 12) |

**Table 6.4** The normalized decision matrix

| | Probability | Exposure | Consequence |
|---|---|---|---|
| H1 | (0.0018, 0.0055, 0.0092, 0.0129, 0.0166, 0.0203) | (0.003, 0.006, 0.0089, 0.0119, 0.0149, 0.0179) | (0.0105, 0.014, 0.0176, 0.0211, 0.0246, 0.0281) |
| H2 | (0.0037, 0.0074, 0.0111, 0.0148, 0.0185, 0.0222) | (0.0089, 0.0119, 0.0149, 0.0179, 0.0209, 0.0238) | (0.0105, 0.014, 0.0176, 0.0211, 0.0246, 0.0281) |
| H3 | (0.0037, 0.0074, 0.0111, 0.0148, 0.0185, 0.0222) | (0.0074, 0.0104, 0.0134, 0.0164, 0.0194, 0.0223) | (0.0105, 0.014, 0.0176, 0.0211, 0.0246, 0.0281) |
| H4 | (0.0055, 0.0092, 0.0129, 0.0166, 0.0203, 0.024) | (0.0089, 0.0119, 0.0149, 0.0179, 0.0209, 0.0238) | (0.0018, 0.0053, 0.0088, 0.0123, 0.0158, 0.0193) |

(continued)

**Table 6.4** (continued)

| | Probability | Exposure | Consequence |
|---|---|---|---|
| H5 | (0.0055, 0.0092, 0.0129, 0.0166, 0.0203, 0.024) | (0.0074, 0.0104, 0.0134, 0.0164, 0.0194, 0.0223) | (0.0053, 0.0088, 0.0123, 0.0158, 0.0193, 0.0228) |
| H6 | (0.0092, 0.0129, 0.0166, 0.0203, 0.024, 0.0277) | (0.0089, 0.0119, 0.0149, 0.0179, 0.0209, 0.0238) | (0.0018, 0.0053, 0.0088, 0.0123, 0.0158, 0.0193) |
| H7 | (0.0037, 0.0074, 0.0111, 0.0148, 0.0185, 0.0222) | (0.0089, 0.0119, 0.0149, 0.0179, 0.0209, 0.0238) | (0.0053, 0.0088, 0.0123, 0.0158, 0.0193, 0.0228) |
| H8 | (0.0055, 0.0092, 0.0129, 0.0166, 0.0203, 0.024) | (0.0089, 0.0119, 0.0149, 0.0179, 0.0209, 0.0238) | (0.0018, 0.0053, 0.0088, 0.0123, 0.0158, 0.0193) |
| H9 | (0.0037, 0.0074, 0.0111, 0.0148, 0.0185, 0.0222) | (0.003, 0.006, 0.0089, 0.0119, 0.0149, 0.0179) | (0.0105, 0.014, 0.0176, 0.0211, 0.0246, 0.0281) |
| H10 | (0.0055, 0.0092, 0.0129, 0.0166, 0.0203, 0.024) | (0.0089, 0.0119, 0.0149, 0.0179, 0.0209, 0.0238) | (0.0035, 0.007, 0.0105, 0.014, 0.0176, 0.0211) |
| H11 | (0.0092, 0.0129, 0.0166, 0.0203, 0.024, 0.0277) | (0.0074, 0.0104, 0.0134, 0.0164, 0.0194, 0.0223) | (0.0035, 0.007, 0.0105, 0.014, 0.0176, 0.0211) |
| H12 | (0.0037, 0.0074, 0.0111, 0.0148, 0.0185, 0.0222) | (0.0074, 0.0104, 0.0134, 0.0164, 0.0194, 0.0223) | (0.0018, 0.0053, 0.0088, 0.0123, 0.0158, 0.0193) |
| H13 | (0.0055, 0.0092, 0.0129, 0.0166, 0.0203, 0.024) | (0.0074, 0.0104, 0.0134, 0.0164, 0.0194, 0.0223) | (0.0035, 0.007, 0.0105, 0.014, 0.0176, 0.0211) |
| H14 | (0.0037, 0.0074, 0.0111, 0.0148, 0.0185, 0.0222) | (0.0089, 0.0119, 0.0149, 0.0179, 0.0209, 0.0238) | (0.0105, 0.014, 0.0176, 0.0211, 0.0246, 0.0281) |
| H15 | (0.0037, 0.0074, 0.0111, 0.0148, 0.0185, 0.0222) | (0.003, 0.006, 0.0089, 0.0119, 0.0149, 0.0179) | (0.0105, 0.014, 0.0176, 0.0211, 0.0246, 0.0281) |
| H16 | (0.0055, 0.0092, 0.0129, 0.0166, 0.0203, 0.024) | (0.0089, 0.0119, 0.0149, 0.0179, 0.0209, 0.0238) | (0.0035, 0.007, 0.0105, 0.014, 0.0176, 0.0211) |
| H17 | (0.0092, 0.0129, 0.0166, 0.0203, 0.024, 0.0277) | (0.0074, 0.0104, 0.0134, 0.0164, 0.0194, 0.0223) | (0.0035, 0.007, 0.0105, 0.014, 0.0176, 0.0211) |
| H18 | (0.0037, 0.0074, 0.0111, 0.0148, 0.0185, 0.0222) | (0.0074, 0.0104, 0.0134, 0.0164, 0.0194, 0.0223) | (0.0018, 0.0053, 0.0088, 0.0123, 0.0158, 0.0193) |
| H19 | (0.0055, 0.0092, 0.0129, 0.0166, 0.0203, 0.024) | (0.0074, 0.0104, 0.0134, 0.0164, 0.0194, 0.0223) | (0.0035, 0.007, 0.0105, 0.014, 0.0176, 0.0211) |
| H20 | (0.0037, 0.0074, 0.0111, 0.0148, 0.0185, 0.0222) | (0.0089, 0.0119, 0.0149, 0.0179, 0.0209, 0.0238) | (0.0105, 0.014, 0.0176, 0.0211, 0.0246, 0.0281) |
| H21 | (0.0037, 0.0074, 0.0111, 0.0148, 0.0185, 0.0222) | (0.003, 0.006, 0.0089, 0.0119, 0.0149, 0.0179) | (0.0105, 0.014, 0.0176, 0.0211, 0.0246, 0.0281) |
| H22 | (0.0055, 0.0092, 0.0129, 0.0166, 0.0203, 0.024) | (0.0089, 0.0119, 0.0149, 0.0179, 0.0209, 0.0238) | (0.0035, 0.007, 0.0105, 0.014, 0.0176, 0.0211) |
| H23 | (0.0092, 0.0129, 0.0166, 0.0203, 0.024, 0.0277) | (0.0074, 0.0104, 0.0134, 0.0164, 0.0194, 0.0223) | (0.0035, 0.007, 0.0105, 0.014, 0.0176, 0.0211) |
| H24 | (0.0037, 0.0074, 0.0111, 0.0148, 0.0185, 0.0222) | (0.0074, 0.0104, 0.0134, 0.0164, 0.0194, 0.0223) | (0.0018, 0.0053, 0.0088, 0.0123, 0.0158, 0.0193) |
| H25 | (0.0037, 0.0074, 0.0111, 0.0148, 0.0185, 0.0222) | (0.0089, 0.0119, 0.0149, 0.0179, 0.0209, 0.0238) | (0.0105, 0.014, 0.0176, 0.0211, 0.0246, 0.0281) |
| H26 | (0.0055, 0.0092, 0.0129, 0.0166, 0.0203, 0.024) | (0.0074, 0.0104, 0.0134, 0.0164, 0.0194, 0.0223) | (0.0035, 0.007, 0.0105, 0.014, 0.0176, 0.0211) |

Then, the ranking of the hazards is obtained for fuzzy ratio system ($FRS_i$). The value of the fuzzy ratio system and the ranking for the fuzzy ratio system are presented in Table 6.5.

At the next step, the fuzzy reference point approach is calculated using the fuzzy ratio system. The value of the fuzzy reference point and the ranking of hazards are presented in Table 6.6.

**Table 6.5** The results of the fuzzy ration system

| | $FRS_i$ | Rank | | $FRS_i$ | Rank |
|---|---|---|---|---|---|
| H1 | 0.0287 | 8 | H14 | 0.0332 | 1 |
| H2 | 0.0332 | 1 | H15 | 0.0297 | 3 |
| H3 | 0.0323 | 2 | H16 | 0.0284 | 9 |
| H4 | 0.0269 | 11 | H17 | 0.0297 | 4 |
| H5 | 0.0290 | 6 | H18 | 0.0250 | 12 |
| H6 | 0.0291 | 5 | H19 | 0.0275 | 10 |
| H7 | 0.0288 | 7 | H20 | 0.0332 | 1 |
| H8 | 0.0269 | 11 | H21 | 0.0297 | 3 |
| H9 | 0.0297 | 3 | H22 | 0.0284 | 9 |
| H10 | 0.0284 | 9 | H23 | 0.0297 | 4 |
| H11 | 0.0297 | 4 | H24 | 0.0250 | 12 |
| H12 | 0.0250 | 12 | H25 | 0.0332 | 1 |
| H13 | 0.0275 | 10 | H26 | 0.0275 | 10 |

**Table 6.6** The results of the fuzzy reference point

| | $r(z_i))$ | Rank | | $r(z_i))$ | Rank |
|---|---|---|---|---|---|
| H1 | 0.00214 | 3 | H14 | 0.00160 | 1 |
| H2 | 0.00160 | 1 | H15 | 0.00175 | 2 |
| H3 | 0.00160 | 1 | H16 | 0.00293 | 5 |
| H4 | 0.00367 | 6 | H17 | 0.00293 | 5 |
| H5 | 0.00220 | 4 | H18 | 0.00367 | 6 |
| H6 | 0.00367 | 6 | H19 | 0.00293 | 5 |
| H7 | 0.00220 | 4 | H20 | 0.00160 | 1 |
| H8 | 0.00367 | 6 | H21 | 0.00175 | 2 |
| H9 | 0.00175 | 2 | H22 | 0.00293 | 5 |
| H10 | 0.00293 | 5 | H23 | 0.00293 | 5 |
| H11 | 0.00293 | 5 | H24 | 0.00367 | 6 |
| H12 | 0.00367 | 6 | H25 | 0.00160 | 1 |
| H13 | 0.00293 | 5 | H26 | 0.00293 | 5 |

Then, the fuzzy full multiplicative form is calculated using the normalized decision matrix. The value of the fuzzy full multiplicative form and the ranking of hazards are presented in Table 6.7.

We used reciprocal rank method at the final step of the fuzzy MULTIMOORA. It considers the position of each alternative with respect to ranking of fuzzy ratio system fuzzy reference point and fuzzy full multiplicative form. The results of the reciprocal rank method are presented in Table 6.8.

**Table 6.7** The results of the fuzzy full multiplicative form

| | $(r(U_i))$ | Rank | | $(r(U_i))$ | Rank |
|---|---|---|---|---|---|
| H1 | 7.01E−06 | 10 | H14 | 1.1266E−05 | 1 |
| H2 | 1.13E−05 | 1 | H15 | 7.86495E−06 | 8 |
| H3 | 1.04E−05 | 2 | H16 | 8.71522E−06 | 6 |
| H4 | 7.76E−06 | 9 | H17 | 9.63312E−06 | 3 |
| H5 | 8.95E−06 | 5 | H18 | 6.5026E−06 | 11 |
| H6 | 9.25E−06 | 4 | H19 | 8.06786E−06 | 7 |
| H7 | 8.72E−06 | 6 | H20 | 1.1266E−05 | 1 |
| H8 | 7.76E−06 | 9 | H21 | 7.86495E−06 | 8 |
| H9 | 7.86E−06 | 8 | H22 | 8.71522E−06 | 6 |
| H10 | 8.72E−06 | 6 | H23 | 9.63312E−06 | 3 |
| H11 | 9.63E−06 | 3 | H24 | 6.5026E−06 | 11 |
| H12 | 6.5E−06 | 11 | H25 | 1.1266E−05 | 1 |
| H13 | 8.07E−06 | 7 | H26 | 8.06786E−06 | 7 |

**Table 6.8** The results of the reciprocal rank method

| | RPA(i) | Rank | | RPA(i) | Rank |
|---|---|---|---|---|---|
| H1 | 1.791 | 7 | H14 | 0.333 | 1 |
| H2 | 0.333 | 1 | H15 | 1.043 | 3 |
| H3 | 0.500 | 2 | H16 | 2.093 | 8 |
| H4 | 2.712 | 10 | H17 | 1.277 | 4 |
| H5 | 1.622 | 5 | H18 | 2.933 | 11 |
| H6 | 1.622 | 5 | H19 | 2.258 | 9 |
| H7 | 1.787 | 6 | H20 | 0.333 | 1 |
| H8 | 2.712 | 10 | H21 | 1.043 | 3 |
| H9 | 1.043 | 3 | H22 | 2.093 | 8 |
| H10 | 2.093 | 8 | H23 | 1.277 | 4 |
| H11 | 1.277 | 4 | H24 | 2.933 | 11 |
| H12 | 2.933 | 11 | H25 | 0.333 | 1 |
| H13 | 2.258 | 9 | H26 | 2.258 | 9 |

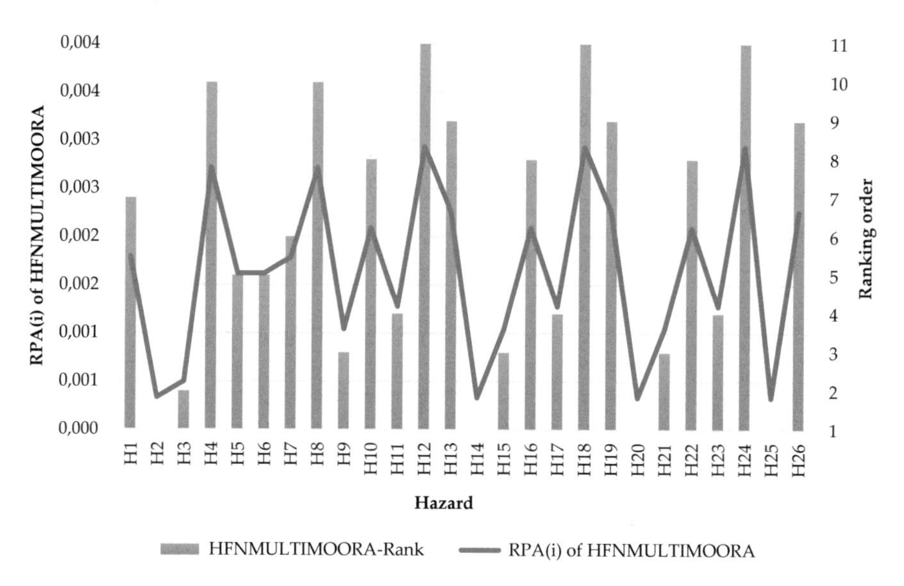


**Fig. 6.2** Final HFMULTIMOORA scores and rankings of each hazard

Graphically, the final HFMULTIMOORA scores and ranking orders of 26 hazards are given in Fig. 6.2. According to these results, the most serious hazards are H2, H14, H20, and H25.

### 6.3.2 *Validation Study on the Results*

To test the validity of the proposed approach, a comparative study is carried on the results. This analysis is about the ranking aggregation tools of the MULTIMOORA. Dominance Theory was used in the original MULTIMOORA method that is based on some principal as dominance, equality and transitiveness. The dominance theory is one of the most used technique in MULTIOORA [27]. In this chapter, the final ranking of the proposed is obtained using the reciprocal rank method. The results of the reciprocal rank method and dominance theory are presented in Table 6.9.

Figure 6.3 and Table 6.9 demonstrate that in both aggregation tools, the most important hazards are H2, H14, H20 and H25. It is also note that, according to a correlation analysis, which measures the association between rankings of hazards, there is a significant and strong positive correlation between the two aggregation tools. The Spearman rank correlation coefficient (RHO) values are higher than 0.980. The Spearman RHO is 0.935 between the classical Fine–Kinney and HFMULTIMOORA using the reciprocal rank method that is significant and strong positive correlation. On the other hand, The Spearman RHO is also significant and strong (0.964) between the classical Fine–Kinney and HFMULTIMOORA using the dominance theory.

**Table 6.9** Results of ranking of hazards with respect to aggregation tools

| Hazards | The classical Fine–Kinney | HFMULTIMOORA (The reciprocal rank method) | HFMULTIMOORA (Dominance theory) |
|---|---|---|---|
| H1 | 8 | 7 | 9 |
| **H2** | **1** | **1** | **1** |
| H3 | 2 | 2 | 2 |
| H4 | 9 | 10 | 12 |
| H5 | 4 | 5 | 7 |
| H6 | 6 | 5 | 6 |
| H7 | 4 | 6 | 8 |
| H8 | 9 | 10 | 12 |
| H9 | 5 | 3 | 4 |
| H10 | 6 | 8 | 10 |
| H11 | 3 | 4 | 3 |
| H12 | 10 | 11 | 13 |
| H13 | 7 | 9 | 11 |
| **H14** | **1** | **1** | **1** |
| H15 | 5 | 3 | 4 |
| H16 | 6 | 8 | 10 |
| H17 | 3 | 4 | 3 |
| H18 | 10 | 11 | 13 |
| H19 | 7 | 9 | 11 |
| **H20** | **1** | **1** | **1** |
| H21 | 5 | 3 | 5 |
| H22 | 6 | 8 | 10 |
| H23 | 3 | 4 | 3 |
| H24 | 10 | 11 | 13 |
| **H25** | **1** | **1** | **1** |
| H26 | 7 | 9 | 11 |

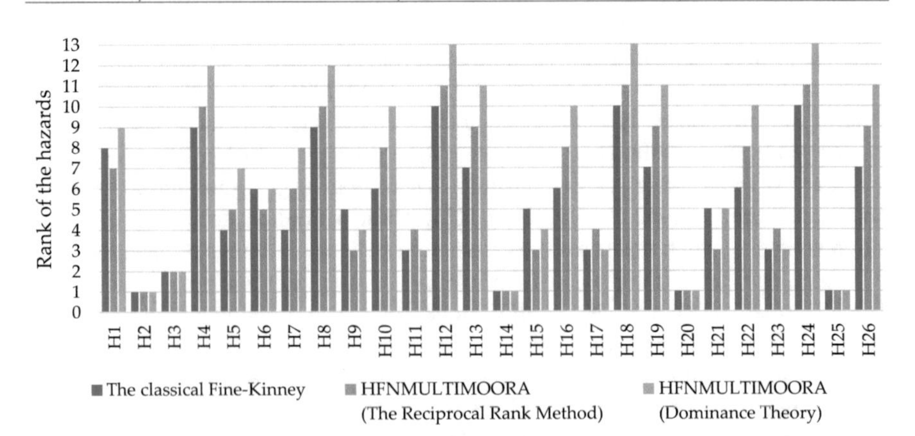


**Fig. 6.3** Results of the aggregation tools

## 6.4 Python Implementation of the Proposed Approach

```
# Chapter 6
# import required libraries
import numpy as np

# set initial variables
n_criteria = 3
n_hazard = 26
nf_score = 6  # number of linguistics
nf_value = 6

criteria_weight = [0.289, 0.293, 0.418]  # P, F, S

linguistic_terms = [  # P, F, S
  ["VL", 0.2, 0.5, 1],
  ["L", 0.5, 1, 3],
  ["ML", 1, 2, 7],
  ["MH", 3, 3, 15],
  ["H", 6, 6, 40],
  ["VH", 10, 10, 100]
]

fuzzy_scores = [["VL", 1, 3, 5, 7, 9, 11],  # Poor (P)
          ["L", 2, 4, 6, 8, 10, 12],  # Medium Poor (MP)
          ["ML", 3, 5, 7, 9, 11, 13],  # Medium (M)
          ["MH", 5, 7, 9, 11, 13, 15],  # Medium Good (MG)
          ["H", 6, 8, 10, 12, 14, 16],  # Good (G)
          ["VH", 7, 9, 11, 13, 15, 17]  # Very Good (VG)
```

```
# ehe: 1st expert's hazard evaluation
ehe = [  # Hazard ID, Probability, Exposure,   Consequence
    ["H1", "VL", "L", "H"],
    ["H2", "L", "H", "H"],
    ["H3", "L", "MH", "H"],
    ["H4", "ML", "H", "VL"],
    ["H5", "ML", "MH", "ML"],
    ["H6", "MH", "H", "VL"],
    ["H7", "L", "H", "ML"],
    ["H8", "ML", "H", "VL"],
    ["H9", "L", "L", "H"],
    ["H10", "ML", "H", "L"],
    ["H11", "MH", "MH", "L"],
    ["H12", "L", "MH", "VL"],
    ["H13", "ML", "MH", "L"],
    ["H14", "L", "H", "H"],
    ["H15", "L", "L", "H"],
    ["H16", "ML", "H", "L"],
    ["H17", "MH", "MH", "L"],
    ["H18", "L", "MH", "VL"],
    ["H19", "ML", "MH", "L"],
    ["H20", "L", "H", "H"],
    ["H21", "L", "L", "H"],
    ["H22", "ML", "H", "L"],
    ["H23", "MH", "MH", "L"],
    ["H24", "L", "MH", "VL"],
    ["H25", "L", "H", "H"],
    ["H26", "ML", "MH", "L"]
]

def rank(decision_vector, direction):  # direction -1:descending, 1:ascending
    order = np.zeros([len(decision_vector), 1])
    unique_val = direction * np.sort(direction * np.unique(decision_vector))
    for ix in range(0, len(unique_val)):
        order[np.argwhere(decision_vector == unique_val[ix])] = ix + 1
    return order
```

```
def print_result(order, vector):
    print('Hazard Id, Rank, Value')
    for ix in range(0, len(order)):
        print(ehe[ix][0], ', ', int(order[ix]), ', ', vector[ix])

idm = [] # initial decision matrix
risk_score = []

# set initial matrices
for hz in ehe:
    temp_list_1 = []
    temp_list_2 = []
    for cr in range(0, n_criteria):
        ix = 0
        for fs in fuzzy_scores:
            if hz[cr + 1] == fs[0]:
                temp_list_1.append(fs[1:])
                temp_list_2.append(linguistic_terms[ix][cr + 1])
            ix += 1
    risk_score.append(np.prod(temp_list_2))
    idm.append(temp_list_1)

denominator = np.sum(np.sqrt(np.sum(np.square(idm), axis=2)), axis=0)
# ndm: normalized decision matrix
# wdm: weighted decision matrix
ndm = np.zeros([n_hazard, n_criteria, nf_value], dtype=float)
wdm = np.zeros([n_hazard, n_criteria, nf_value], dtype=float)
for hz_ix in range(0, n_hazard):
    for cr_ix in range(0, n_criteria):
        for fs_ix in range(0, nf_value):
            ndm[hz_ix][cr_ix][fs_ix] = idm[hz_ix][cr_ix][fs_ix] / denominator[cr_ix]
            wdm[hz_ix][cr_ix][fs_ix] = criteria_weight[cr_ix] * ndm[hz_ix][cr_ix][fs_ix]

temp_val = [0.25, 0.5, 0.25, 0.25, 0.5, 0.25] # [1/4, 2/4, 1/4, 1/4, 2/4,1/4]
y = np.sum(np.sum(wdm, axis=1) * temp_val, axis=1) # fuzzy ratio
X_plus = np.max(wdm, axis=0)
reference_point = np.sum(np.abs(wdm - X_plus), axis=2) / 6
z = np.max(reference_point, axis=1)
multiplicative_form = np.asarray(np.prod(ndm, axis=1))
u = np.sum(multiplicative_form * temp_val, axis=1)
order_y, order_z, order_u = rank(y, -1), rank(z, 1), rank(np.round(u, 20), -1)
```

```
RPA = 1 / (1 / order_y + 1 / order_z + 1 / order_u)

hazard_rank = rank(RPA, 1)
print_result(hazard_rank, RPA)
'''
Output:
Hazard Id, Rank, Value
H1 , 11 , [1.79104478]
H2 , 1 , [0.33333333]
H3 , 2 , [0.5]
H4 , 10 , [2.71232877]
H5 , 5 , [1.62162162]
H6 , 5 , [1.62162162]
H7 , 6 , [1.78723404]
H8 , 10 , [2.71232877]
H9 , 3 , [1.04347826]
H10 , 8 , [2.09302326]
H11 , 4 , [1.27659574]
H12 , 11 , [2.93333333]
H13 , 9 , [2.25806452]
H14 , 1 , [0.33333333]
H15 , 3 , [1.04347826]
H16 , 8 , [2.09302326]
H17 , 4 , [1.27659574]
H18 , 11 , [2.93333333]
H19 , 9 , [2.25806452]
H20 , 1 , [0.33333333]
H21 , 3 , [1.04347826]
H22 , 8 , [2.09302326]
H23 , 4 , [1.27659574]
H24 , 11 , [2.93333333]
H25 , 1 , [0.33333333]
H26 , 9 , [2.25806452]
'''
```

## References


1. Rajarajeswari, P., Sudha, A. S., & Karthika, R. (2013). A new operation on hexagonal fuzzy number. *International Journal of Fuzzy Logic Systems, 3*(3), 15–26.
2. Dinagar, D. S., & Narayanan, U. H. (2016). On determinant of hexagonal fuzzy number matrices. *International Journal of Mathematics And its Applications, 4*(4), 357–363.
3. Rajarajeswari, P., & Sudha, A. S. (2014). Ordering generalized hexagonal fuzzy numbers using rank, mode, divergence and spread. *IOSR Journal of Mathematics, 10*(3), 15–22.
4. Chakraborty, A., Maity, S., Jain, S., Mondal, S. P., & Alam, S. (2020). Hexagonal fuzzy number and its distinctive representation, ranking, defuzzification technique and application in production inventory management problem. *Granular Computing*, 1–15.
5. Deshmukh, M. C., Ghadle, K. P., & Jadhav, O. S. (2020). Optimal solution of fully fuzzy LPP with symmetric HFNs. In *Computing in engineering and technology* (pp. 387–395). Singapore: Springer.
6. Parveen, N., & Kamble, P. N. (2020). Decision-making problem using fuzzy TOPSIS Method with hexagonal fuzzy number. In *Computing in engineering and technology* (pp. 421–430). Singapore: Springer.
7. Brauers, W. K., & Zavadskas, E. K. (2006). The MOORA method and its application to privatization in a transition economy. *Control and Cybernetics, 35,* 445–469.
8. Hafezalkotob, A., Hafezalkotob, A., Liao, H., & Herrera, F. (2019). An overview of MULTIMOORA for multi-criteria decision-making: Theory, developments, applications, and challenges. *Information Fusion, 51,* 145–177.
9. Deliktas, D., & Ustun, O. (2017). Student selection and assignment methodology based on fuzzy MULTIMOORA and multichoice goal programming. *International Transactions in Operational Research, 24*(5), 1173–1195.
10. Mavi, R. K., Goh, M., & Zarbakhshnia, N. (2017). Sustainable third-party reverse logistic provider selection with fuzzy SWARA and fuzzy MOORA in plastic industry. *The International Journal of Advanced Manufacturing Technology, 91*(5–8), 2401–2418.
11. Baležentis, A., Baležentis, T., & Brauers, W. K. (2012). Personnel selection based on computing with words and fuzzy MULTIMOORA. *Expert Systems with Applications, 39*(9), 7961–7967.
12. Baležentis, T., & Zeng, S. (2013). Group multi-criteria decision making based upon interval-valued fuzzy numbers: an extension of the MULTIMOORA method. *Expert Systems with Applications, 40*(2), 543–550.
13. Stanujkic, D., Zavadskas, E. K., Smarandache, F., Brauers, W. K., & Karabasevic, D. (2017). A neutrosophic extension of the MULTIMOORA method. *Informatica, 28*(1), 181–192.
14. Dorfeshan, Y., Mousavi, S. M., Mohagheghi, V., & Vahdani, B. (2018). Selecting project-critical path by a new interval type-2 fuzzy decision methodology based on MULTIMOORA, MOOSRA and TPOP methods. *Computers & Industrial Engineering, 120,* 160–178.
15. Liu, H. C., You, J. X., Lu, C., & Shan, M. M. (2014). Application of interval 2-tuple linguistic MULTIMOORA method for health-care waste treatment technology evaluation and selection. *Waste Management, 34*(11), 2355–2364.
16. Gou, X., Liao, H., Xu, Z., & Herrera, F. (2017). Double hierarchy hesitant fuzzy linguistic term set and MULTIMOORA method: A case of study to evaluate the implementation status of haze controlling measures. *Information Fusion, 38,* 22–34.
17. Liao, H., Qin, R., Gao, C., Wu, X., Hafezalkotob, A., & Herrera, F. (2019). Score-HeDLiSF: A score function of hesitant fuzzy linguistic term set based on hesitant degrees and linguistic scale functions: An application to unbalanced hesitant fuzzy linguistic MULTIMOORA. *Information Fusion, 48,* 39–54.
18. Zhang, C., Chen, C., Streimikiene, D., & Balezentis, T. (2019). Intuitionistic fuzzy MULTI-MOORA approach for multi-criteria assessment of the energy storage technologies. *Applied Soft Computing, 79,* 410–423.
19. Mete, S. (2019). Assessing occupational risks in pipeline construction using FMEA-based AHP-MOORA integrated approach under Pythagorean fuzzy environment. *Human and Ecological Risk Assessment: An International Journal, 25*(7), 1645–1660.

20. Liang, D., Darko, A. P., & Zeng, J. (2019). Interval-valued pythagorean fuzzy power average-based MULTIMOORA method for multi-criteria decision-making. *Journal of Experimental & Theoretical Artificial Intelligence*, 1–30.
21. Büyüközkan, G., & Göçer, F. (2017, June). An extension of MOORA approach for group decision making based on interval valued intuitionistic fuzzy numbers in digital supply chain. In *2017 Joint 17th world congress of international fuzzy systems association and 9th international conference on soft computing and intelligent systems (IFSA-SCIS)* (pp. 1–6). IEEE.
22. Baležentis, T., & Baležentis, A. (2014). A survey on development and applications of the multi-criteria decision making method MULTIMOORA. *Journal of Multi-Criteria Decision Analysis, 21*(3–4), 209–222.
23. Dhurai, K., & Karpagam, A. (2016). A new pivotal operation on triangular fuzzy number for solving fully fuzzy linear programming problems. *International Journal of Applied Mathematical Sciences, 9*(1), 41–46.
24. Altuntas, S., Dereli, T. & Yilmaz, M. K. (2015). Evaluation of excavator technologies: Application of data fusion based MULTIMOORA methods. *Journal of Civil Engineering and Management, 21*(8), 977–997.
25. Şardan, H. S. (2005). İş sağlığı ve güvenliğinde yeni oluşumlar; risk değerlendirmesi ve OHSAS 18001. Çimento Müstahsilleri İşverenleri Sendikası.
26. Gul, M., Guven, B., & Guneri, A. F. (2018). A new Fine–Kinney-based risk assessment framework using FAHP-FVIKOR incorporation. *Journal of Loss Prevention in the Process Industries, 53*, 3–16.
27. Brauers, W. K. M., & Zavadskas, E. K. (2012). Robustness of MULTIMOORA: A method for multi-objective optimization. *Informatica, 23*(1), 1–25.

# Chapter 7
# Fine–Kinney-Based Occupational Risk Assessment Using Single-Valued Neutrosophic TOPSIS

**Abstract** Neutrosophic sets are initially recommended by Smarandache (First international conference on neutrosophy, neutrosophic logic, set, probability, and statistics. University of New Mexico, Gallup, NM, pp 338–353, 2002 [1]). These sets reflect uncertainty and vagueness in real-world problems better than classical fuzzy set theory. It takes into consideration three decision-making situations called indeterminacy, truthiness, and falsity. In Zadeh traditional fuzzy set theory, there is just membership function fuzzy set degree. But, in neutrosophic environment, it considers three membership functions. Unlike intuitionistic fuzzy sets, an indeterminacy degree is considered. In this chapter, we applied a special form of neutrosophic set as single-valued neutrosophic set (SVNs) with the technique for order preference by similarity to ideal solution (TOPSIS) under the concept of Fine–Kinney occupational risk assessment. Since the mere TOPSIS has failed to handle imprecise and vague information which usually exist in real-world problems, we follow the integration of SVNs and TOPSIS. To demonstrate the applicability of the novel approach, a case study of risk assessment of a wind turbine in times of operation was provided. Comparative analysis with some similar approaches and sensitivity analysis by changing the weights of Fine–Kinney parameters are carried out. Finally, the Python implementation of the proposed approach is executed to be useful for those concerned in the future.

## 7.1 Neutrosophic Sets and TOPSIS

In this part, before explaining the proposed approach, we give some preliminaries on neutrosophic sets and more specifically single-valued neutrosophic sets. Then, the algorithm of single-valued neutrosophic technique for order preference by similarity to ideal solution (SVNTOPSIS) is given in detail.



M. Gul et al., *Fine–Kinney-Based Fuzzy Multi-criteria Occupational Risk Assessment*, Studies in Fuzziness and Soft Computing 398,
https://doi.org/10.1007/978-3-030-52148-6_7

### 7.1.1 General View on Single-Valued Neutrosophic Sets

Single-valued neutrosophic sets is a particular version of neutrosophic sets. Here, some definitions, operations, and distance measurement between two single-valued neutrosophic sets are provided [1–4].

**Definition 1** A single-valued neutrosophic set is demonstrated by three different membership functions as follows [2–4]:

- The truth-membership function $T_{\tilde{n}}(x)$,
- The indeterminacy-membership function $I_{\tilde{n}}(x)$, and
- The falsity-membership function $F_{\tilde{n}}(x)$. For all $x$, $T_{\tilde{n}}(x), I_{\tilde{n}}(x), F_{\tilde{n}}(x) \epsilon [0, 1]$.

For the sum of three membership functions of a single-valued neutrosophic set, the relation of $0 \leq T_{\tilde{n}}(x) + I_{\tilde{n}}(x) + F_{\tilde{n}}(x) \leq 3$ is valid for all $x$.

**Definition 2** Mathematical operations between two single-valued neutrosophic sets are determined as shown below [2–4]

Let $\tilde{n} = \langle T_{\tilde{n}}(x), I_{\tilde{n}}(x), F_{\tilde{n}}(x) \rangle$ and $\tilde{s} = \langle T_{\tilde{s}}(x), I_{\tilde{s}}(x), F_{\tilde{s}}(x) \rangle$, be two single-valued neutrosophic sets. The addition, multiplication, union, and intersection operations can be calculated as in Eqs. (7.1–7.4).

$$\tilde{n} \oplus \tilde{s} = \langle T_{\tilde{n}}(x) + T_{\tilde{s}}(x) - T_{\tilde{n}}(x) * T_{\tilde{s}}(x), I_{\tilde{n}}(x) * I_{\tilde{s}}(x), F_{\tilde{s}}(x) * F_{\tilde{s}}(x) \rangle \text{ for all } x \tag{7.1}$$

$$\tilde{n} \otimes \tilde{s} = \langle T_{\tilde{n}}(x) * T_{\tilde{s}}(x), I_{\tilde{n}}(x) + I_{\tilde{s}}(x) - I_{\tilde{n}}(x) * I_{\tilde{s}}(x), F_{\tilde{n}}(x) + F_{\tilde{s}}(x) - F_{\tilde{n}}(x) * F_{\tilde{s}}(x) \rangle \tag{7.2}$$

$$\tilde{n} \cup \tilde{s} = \langle \max(T_{\tilde{n}}(x), T_{\tilde{s}}(x)), \min(I_{\tilde{n}}(x), I_{\tilde{s}}(x)), \min(F_{\tilde{n}}(x), F_{\tilde{s}}(x)) \rangle \text{ for all } x \tag{7.3}$$

$$\tilde{n} \cap \tilde{s} = \langle \min(T_{\tilde{n}}(x), T_{\tilde{s}}(x)), \max(I_{\tilde{n}}(x), I_{\tilde{s}}(x)), \max(F_{\tilde{n}}(x), F_{\tilde{s}}(x)) \rangle \text{ for all } x \tag{7.4}$$

**Definition 3** Euclidean distance of between two single-valued neutrosophic sets is calculated as follows [2–4]:

Let $\tilde{n} = \{(x_1 | \langle T_{\tilde{n}}(x_1), I_{\tilde{n}}(x_1), F_{\tilde{n}}(x_1) \rangle), \ldots, (x_m | \langle T_{\tilde{n}}(x_m), I_{\tilde{n}}(x_m), F_{\tilde{n}}(x_m) \rangle) \}$ and $\tilde{s} = \{(x_1 | \langle T_{\tilde{s}}(x_1), I_{\tilde{s}}(x_1), F_{\tilde{s}}(x_1) \rangle), \ldots, (x_m | \langle T_{\tilde{s}}(x_m), I_{\tilde{s}}(x_m), F_{\tilde{s}}(x_m) \rangle) \}$ be two single-valued neutrosophic sets for $x_i \in X (i = 1, 2, \ldots, m)$. The Euclidean distance and normalized Euclidean distance between these two sets are computed as in Eqs. (7.5–7.6).

$$d\_\text{Eucl}(\tilde{n}, \tilde{s}) = \sqrt{\sum_{i=1}^{m} \{(T_{\tilde{n}}(x_i) - T_{\tilde{s}}(x_i))^2 + (I_{\tilde{n}}(x_i) - I_{\tilde{s}}(x_i))^2 + (F_{\tilde{n}}(x_i) - F_{\tilde{s}}(x_i))^2\}} \tag{7.5}$$

$$\text{distance\_normalized\_Eucl}(\tilde{n}, \tilde{s}) = \sqrt{\frac{1}{3m} * \sum_{i=1}^{m}\{(T_{\tilde{n}}(x_i) - T_{\tilde{s}}(x_i))^2 + (I_{\tilde{n}}(x_i) - I_{\tilde{s}}(x_i))^2 + (F_{\tilde{n}}(x_i) - F_{\tilde{s}}(x_i))^2\}} \tag{7.6}$$

### *7.1.2 TOPSIS Method*

The procedural steps of the base TOPSIS which were initially proposed by [5], are summarized as follows:

1. Determine the main goal of the problem and build a decision matrix that covers criteria, alternatives, and Performance measures of alternatives according to criteria,
2. Normalize the decision matrix, which is provided in Step 1,
3. Construct a weighted normalized decision matrix considering a weight matrix of criteria obtained before by a separate method,
4. Define the ideal and the anti-ideal solution,
5. Compute distances from the ideal and the anti-ideal solutions,
6. Compute the relative distance of an alternative to the ideal solution, and
7. Sort the alternatives in descending order.

## 7.2 Proposed Fine–Kinney-Based Approach Using SVNTOPSIS

The base TOPSIS method was integrated with fuzzy sets and its extensions like triangular fuzzy sets [6], interval-valued fuzzy sets [7], neutrosophic sets [2], interval type-2 fuzzy sets [8, 9], hesitant fuzzy sets [10], intuitionistic fuzzy sets [11], and Pythagorean fuzzy sets [12–18]. The detailed procedural flow of SVNTOPSIS is as follows:

**Step 1**: First step generates the decision matrix under single-valued neutrosophic environment [2]. It is denoted with $D_{\tilde{n}} = \left\langle d_{ij}^{s} \right\rangle_{m \times n} = \left\langle T_{ij}, I_{ij}, F_{ij} \right\rangle_{m \times n}$. Its matrix notation is as in Eq. (7.7).

$$\begin{array}{c} \\ A_1 \\ A_2 \\ \cdots \\ A_m \end{array} \begin{array}{c} \begin{array}{cccc} C_1 & C_2 & \cdots & C_n \end{array} \\ \begin{pmatrix} \langle T_{11}, I_{11}, F_{11} \rangle & \langle T_{12}, I_{12}, F_{12} \rangle & \cdots & \langle T_{1n}, I_{1n}, F_{1n} \rangle \\ \langle T_{21}, I_{21}, F_{21} \rangle & \langle T_{22}, I_{22}, F_{22} \rangle & \cdots & \langle T_{2n}, I_{2n}, F_{2n} \rangle \\ \cdots & \cdots & \cdots & \cdots \\ \langle T_{m1}, I_{m1}, F_{m1} \rangle & \langle T_{m2}, I_{m2}, F_{m2} \rangle & \cdots & \langle T_{mn}, I_{mn}, F_{mn} \rangle \end{pmatrix} \end{array} \tag{7.7}$$

Here, $\langle T_{ij}, I_{ij}, F_{ij} \rangle_{m \times n}$ refers to the degree of truth, indeterminacy and falsity-membership value of alternative $A_i$ according to criterion $C_j$ ($i = 1, 2, \ldots, m$ and $j = 1, 2, \ldots, n$).

**Step 2**: The second step concerns with the determination of the weights of decision-makers (OHS experts). Let $E_k = T_k, I_k, F_k$ be a neutrosophic number for the evaluation of $k$th OHS expert. The weight of this OHS expert can be computed by using Eq. (7.8) [2]

$$w_k(\text{weight of } k\text{th OHS expert}) = \frac{1 - \sqrt{\left\{\left(1 - T_k(x)^2\right) + \left(I_k(x)^2\right) + \left(F_k(x)^2\right)\right\}/3}}{\sum_{k=1}^{p}\left(1 - \sqrt{\frac{\left\{\left(1 - T_k(x)^2\right) + \left(I_k(x)^2\right) + \left(F_k(x)^2\right)\right\}}{3}}\right)} \text{ and } \sum_{k=1}^{p}(w_k) = 1 \tag{7.8}$$

**Step 3**: The third step builds the aggregated decision matrix under single-valued neutrosophic sets depend on experts' evaluations. This matrix is calculated by single-valued neutrosophic weighted averaging operator (SVNWAO) as shown below in Eq. (7.9) [2]

$$\text{SVNWAO}_w\left(d_{ij}^1, d_{ij}^2, \ldots, d_{ij}^r\right) = w_1 d_{ij}^1 \oplus w_2 d_{ij}^2 \oplus \ldots \oplus w_k d_{ij}^r = \left\langle 1 - \prod_{p=1}^{r}\left(1 - T_{ij}^r\right)^{w_k}, \prod_{p=1}^{r}\left(I_{ij}^r\right)^{w_k}, \prod_{p=1}^{r}\left(F_{ij}^r\right)^{w_k} \right\rangle \tag{7.9}$$

**Step 4**: In this step, criteria weights are determined using any other MCDM method such as AHP, BWM or an adapted SVNWAO as in Eq. (7.10).

Here $\omega_j^1, \omega_j^2, \ldots, \omega_j^r$ refers to the neutrosophic matrix assigned to the criterion $C_j$ by the $k$th OHS expert.

$$\text{SVNWAO}_w\left(\omega_j^1, \omega_j^2, \ldots, \omega_j^r\right) = w_1 \omega_j^1 \oplus w_2 \omega_j^2 \oplus \ldots \oplus w_k \omega_j^r = \left\langle 1 - \prod_{p=1}^{r}\left(1 - T_j^r\right)^{w_k}, \prod_{p=1}^{r}\left(I_j^r\right)^{w_k}, \prod_{p=1}^{r}\left(F_j^r\right)^{w_k} \right\rangle \tag{7.10}$$

**Step 5**: In this step, the aggregated weighted single-valued neutrosophic matrix is computed as in Eq. (7.11).

$$D_{\tilde{n}}^{\omega^j} = \left\langle d_{ij}^{\omega^j} \right\rangle_{m \times n} = \left\langle T_{ij}^{\omega^j}, I_{ij}^{\omega^j}, F_{ij}^{\omega^j} \right\rangle_{m \times n} \tag{7.11}$$

**Step 6**: This step covers the determination of relative neutrosophic positive and negative ideal solution shorten as (RNPIS), (RNNIS), respectively, given in Eqs. (7.12–7.13).

$$\text{RNPIS}_{\tilde{n}}^{+} = \left[d_1^{\omega^+}, d_2^{\omega^+}, \ldots, d_n^{\omega^+}\right] \quad (7.12)$$

where $d_j^{\omega^+} = T_j^{\omega^+}, I_j^{\omega^+}, F_j^{\omega^+}$ for $j = 1, 2, \ldots, n$.

$$d_j^{\omega^+} = \left\langle \begin{array}{l} \left\{\left(\max_i\{T_{ij}^{\omega^j}\}\middle| j \in J_1\right), \left(\min_i\{T_{ij}^{\omega^j}\}\middle| j \in J_2\right)\right\}, \left\{\left(\min_i\{I_{ij}^{\omega^j}\}\middle| j \in J_1\right), \left(\max_i\{I_{ij}^{\omega^j}\}\middle| j \in J_2\right)\right\}, \\ \left\{\left(\min_i\{F_{ij}^{\omega^j}\}\middle| j \in J_1\right), \left(\max_i\{F_{ij}^{\omega^j}\}\middle| j \in J_2\right)\right\} \end{array} \right\rangle$$

$$\text{RNNIS}_{\tilde{n}}^{-} = \left[d_1^{\omega^-}, d_2^{\omega^-}, \ldots, d_n^{\omega^-}\right] \quad (7.13)$$

where $d_j^{\omega^-} = \left\langle T_j^{\omega^-}, I_j^{\omega^-}, F_j^{\omega^-} \right\rangle$ for $j = 1, 2, \ldots, n$.

$$d_j^{\omega^-} = \left\langle \begin{array}{l} \left\{\left(\min_i\{T_{ij}^{\omega^j}\}\middle| j \in J_1\right), \left(\max_i\{T_{ij}^{\omega^j}\}\middle| j \in J_2\right)\right\}, \left\{\left(\max_i\{I_{ij}^{\omega^j}\}\middle| j \in J_1\right), \left(\min_i\{I_{ij}^{\omega^j}\}\middle| j \in J_2\right)\right\}, \\ \left\{\left(\max_i\{F_{ij}^{\omega^j}\}\middle| j \in J_1\right), \left(\min_i\{F_{ij}^{\omega^j}\}\middle| j \in J_2\right)\right\} \end{array} \right\rangle$$

**Step 7**: Following the sixth step, the seventh step considers the determination of distances from RNPIS and RNNIS as provided in Eqs. (7.14–7.15). While Eq. (7.14), shows the normalized Euclidean distance from RNPIS, Eq. (7.15), shows the normalized Euclidean distance from RNNIS.

$$d\_\text{Eucl}^{+}(d_{ij}^{\omega^j}, d_j^{\omega^+}) = \sqrt{\frac{1}{3n}\sum_{j=1}^{n}\left\{\left(T_{ij}^{\omega^j}(x_j) - T_j^{\omega^+}(x_j)\right)^2 + \left(I_{ij}^{\omega^j}(x_j) - I_j^{\omega^+}(x_j)\right)^2 + \left(F_{ij}^{\omega^j}(x_j) - F_j^{\omega^+}(x_j)\right)^2\right\}} \quad (7.14)$$

$$d\_\text{Eucl}^{-}(d_{ij}^{\omega^j}, d_j^{\omega^-}) = \sqrt{\frac{1}{3n}\sum_{j=1}^{n}\left\{\left(T_{ij}^{\omega^j}(x_j) - T_j^{\omega^-}(x_j)\right)^2 + \left(I_{ij}^{\omega^j}(x_j) - I_j^{\omega^-}(x_j)\right)^2 + \left(F_{ij}^{\omega^j}(x_j) - F_j^{\omega^-}(x_j)\right)^2\right\}} \quad (7.15)$$

**Step 8**: The last step determines the relative closeness coefficient as shown in Eq. (7.16), and ranking orders of alternatives. The higher closeness coefficient value the better alternative.

$$C_i^{*} = \frac{d\_\text{Eucl}^{-}(d_{ij}^{\omega^j}, d_j^{\omega^-})}{d_{\text{Eucl}}^{+}\left(d_{ij}^{\omega^j}, d_j^{\omega^+}\right) + d\_\text{Eucl}^{-}(d_{ij}^{\omega^j}, d_j^{\omega^-})} \quad (7.16)$$

Since the neutrosophic set is a common version of classical, fuzzy, and intuitionistic fuzzy sets, it reflects uncertainty well and illustrates real-world problems with

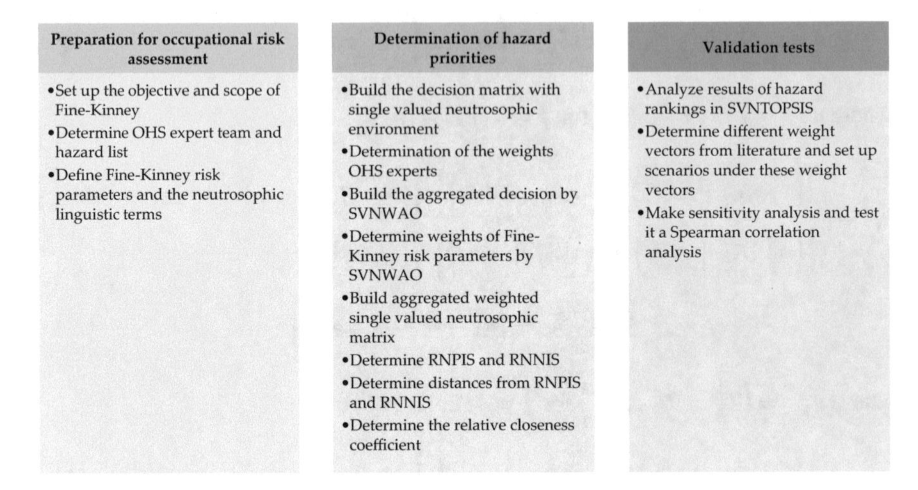


**Fig. 7.1** The step of the proposed approach

its three membership degree in decision-making as truthiness, indeterminacy, and falsity. Therefore, integration of this set with TOPSIS handles vagueness and uncertainty over fuzzy TOPSIS and intuitionistic fuzzy TOPSIS. In this chapter, a novel Fine–Kinney-based SVNTOPSIS for occupational risk assessment is proposed.

Firstly, a group of OHS experts (three OHS experts) determines the objective of the study and scope of the approach. Then, they identify the hazards and determine risk parameters and evaluation scale in neutrosophic environment.

The proposed approach main steps are graphically demonstrated in Fig. 7.1. There are three main steps and their sub-steps through the study. The first is regarding preparation for occupational risk assessment. Second is about OHS experts' linguistic evaluations of each hazard according to three risk parameters of Fine–Kinney concept. They are summed to obtain a mean value. Then, using the linguistic terms, hazards are prioritized using SVNTOPSIS. Finally, some validation tests are performed.

## 7.3 Case Study

To show the proposed approach applicability, a case study for occupational risk assessment of wind turbine during operation period is carried out. The implementation of proposed SVNTOPSIS under Fine–Kinney approach is given in detail in the following subsections.

### 7.3.1 Application Results

In this case study, thirty-one hazards and their associated risks are evaluated by the proposed approach with respect to the three Fine–Kinney parameters. Three experts (E-1, E-2, and E-3) with different experience levels in wind energy industry performed the evaluations. The hazards which are indicated by "H" are adapted from [19]. The most important hazard sources and risks described by OHS experts in the monitored wind turbine are given in Table 7.1.

**Table 7.1** Hazard list in the observed wind turbine during operation period (Reprinted from Ref. [19], Copyright 2018, with permission from Global Journal of Environmental Science and Management (GJESM))

| Hazard | Hazard identification | Associated risk definition |
|---|---|---|
| H1 | Lockers | Fall risk of lockers |
| H2 | Fire | Risk of fire |
| H3 | Dress cabinet | Fall of dress cabinet |
| H4 | Using stairs | Wet floor |
| H5 | Internal transformer | Explosion risk of internal transformer |
| H6 | Human factor | Entry of unauthorized persons to the areas where diesel generator and internal transformer are placed |
| H7 | Septic and water tank | Drowning |
| H8 | Pests and insects | Pest and insect bites |
| H9 | Spraying engaged staff | Electric shock |
| H10 | Electric shock possibility | Possibility of electric shock of security personal |
| H11 | Access of unauthorized persons to waste containers | Poisoning as a result of contact of unauthorized persons to chemicals |
| H12 | Access of unauthorized persons to storage area | Aimless movement of unauthorized persons in warehouse and waste area |
| H13 | Access of unauthorized persons to storage area | Touching and climbing of unauthorized persons to the high voltage towers |
| H14 | Agriculture in the agricultural lands of operational area | Electric shock as a result of plowing the fields and excavations by farmers in the cable route |
| H15 | Opening of the water trenches on the roadside and studies with work machine in the operational area | Electric shock by contacting the MV cables during the works on opening of the water trenches on the roadside and with work machines |
| H16 | Damaging of the heavy rainfall to trench channel | Damage risk of cables as a result of disclosure of the trench channels due to heavy rainfall |

(continued)

**Table 7.1** (continued)

| Hazard | Hazard identification | Associated risk definition |
|---|---|---|
| H17 | Lightning, Ice fall, Overthrow of turbines as a result of the natural disasters | Lightning, Wounding risk as a result of skidding down of ice blocks when moving of iced tower,<br>Wound or death risk as a result of overthrow of wind turbines during natural disasters |
| H18 | The entry of unauthorized persons | Exposure to electric current as a result of entry of unauthorized persons |
| H19 | Works in the turbine area | Entering of unauthorized persons to the turbine working areas |
| H20 | Transformer | High temperature and pressure that may occur in the transformer |
| H21 | Transformer | Spreading of oil as a result of explosion |
| H22 | Transformer | The entry of unauthorized persons |
| H23 | Transformer | Accident resulting in material damage and spreading |
| H24 | RMU cell | Exposure to electric current, Explosion burns |
| H25 | RMU cell | The arcing in the explosion during the maneuver |
| H26 | RMU cell | The entry of unauthorized persons |
| H27 | RMU cell | Low voltage electric shock during operation and intervene in the control panel |
| H28 | Concrete kiosk | Damages of insects and rodents to the cable systems |
| H29 | Concrete kiosk | Entry of unauthorized persons |
| H30 | Concrete kiosk | Damage as a result of fire |
| H31 | Rectifiers | Exposure to electric current |

In the second main step, initially, weights of OHS experts are determined using Eq. (7.8). Considering their years of experience, linguistic terms are assigned to them. The linguistic terms utilized for ratings of OHS experts are given in Table 7.2.

**Table 7.2** Linguistic relation terms and their respected SVN numbers. Reprinted from Ref. [2], with kind permission from Springer Science + Business Media

| Linguistic term | SVN number | | |
|---|---|---|---|
| Very important (VI) | 0.90 | 0.10 | 0.10 |
| Important (I) | 0.80 | 0.20 | 0.15 |
| Medium (M) | 0.50 | 0.40 | 0.45 |
| Unimportant (U) | 0.35 | 0.60 | 0.70 |
| Very unimportant (VU) | 0.10 | 0.80 | 0.90 |

**Table 7.3** Linguistic terms and their associated SVN numbers. Reprinted from ref. [2], with kind permission from Springer Science + Business Media

| Linguistic term | SVN number | | |
|---|---|---|---|
| Extremely high (EH) | 1.00 | 0.00 | 0.00 |
| Very high (VH) | 0.90 | 0.10 | 0.05 |
| High (H) | 0.80 | 0.20 | 0.15 |
| Medium high (MH) | 0.65 | 0.35 | 0.30 |
| Medium (M) | 0.50 | 0.50 | 0.45 |
| Medium low (ML) | 0.35 | 0.65 | 0.60 |
| Low (L) | 0.20 | 0.75 | 0.80 |
| Very low (VL) | 0.10 | 0.85 | 0.90 |
| Extremely low (EL) | 0.05 | 0.90 | 0.95 |

VI, I, and M are given to each OHS expert in SVN environment, respectively. Then, the weight of each OHS expert is computed using Eq. (7.8). The assigned weight values are as follows: $w_{E-1} = 0.436$, $w_{E-2} = 0.395$, $w_{E-3} = 0.169$.

After determination of experts' weights, the aggregated neutrosophic decision matrix is obtained. During these sub-steps, the linguistic terms presented by [2], are used as in Table 7.3.

The aggregated neutrosophic matrix is obtained using Eq. (7.9). Here, three experts' evaluations are considered. The obtained values of hazards according to three Fine–Kinney risk parameters are given in Table 7.4, in neutrosophic style.

Hereafter the determination of aggregated SVN matrix, the weight of each Fine–Kinney risk parameter for each expert, is determined using the linguistic terms in Table 7.2. In this calculation, the formulae of SVNWAO as in Eq. (7.10), is used. Evaluations of risk parameter weights for three experts are given in Table 7.5, as well as the final weight values in SVN style are obtained from SVNWAO formulation.

Following this sub-step, aggregated weighted single-valued neutrosophic matrix is computed using Eq. (7.11). The aggregated weighted SVN values of hazards according to probability, exposure and consequence are shown in Table 7.6.

Then RNPIS and RNNIS values are calculated using Eqs. (7.12–7.13). The achieved results are given in Table 7.7.

Finally, the distance measures and closeness coefficient of SVNTOPSIS are computed using Eqs. (7.14–7.15). The results regarding these computations and ranking of hazards are shown in Table 7.8.

Graphically, the final SVNTOPSIS scores and ranking orders of 31 hazards are given in Fig. 7.2. According to these results, the most serious hazards are H7, H2, H5, H14, and H9.

**Table 7.4** Aggregated SVN values of hazards with respect to Fine–Kinney risk parameters

| Hazard | Probability | | | Exposure | | | Consequence | | |
|---|---|---|---|---|---|---|---|---|---|
| H1 | 0.800 | 0.200 | 0.150 | 0.865 | 0.135 | 0.081 | 0.200 | 0.750 | 0.800 |
| H2 | 0.900 | 0.100 | 0.050 | 0.200 | 0.750 | 0.800 | 0.888 | 0.112 | 0.060 |
| H3 | 0.650 | 0.350 | 0.300 | 0.200 | 0.750 | 0.800 | 0.200 | 0.750 | 0.800 |
| H4 | 0.650 | 0.350 | 0.300 | 0.200 | 0.750 | 0.800 | 0.650 | 0.350 | 0.300 |
| H5 | 0.900 | 0.100 | 0.050 | 0.092 | 0.858 | 0.908 | 1.000 | 0.000 | 0.000 |
| H6 | 0.800 | 0.200 | 0.150 | 0.081 | 0.869 | 0.919 | 0.822 | 0.178 | 0.125 |
| H7 | 0.800 | 0.200 | 0.150 | 0.650 | 0.350 | 0.300 | 0.822 | 0.178 | 0.125 |
| H8 | 0.650 | 0.350 | 0.300 | 0.414 | 0.586 | 0.536 | 0.200 | 0.750 | 0.800 |
| H9 | 0.378 | 0.622 | 0.572 | 0.445 | 0.555 | 0.504 | 0.888 | 0.112 | 0.060 |
| H10 | 0.800 | 0.200 | 0.150 | 0.200 | 0.750 | 0.800 | 0.650 | 0.350 | 0.300 |
| H11 | 0.420 | 0.580 | 0.529 | 0.059 | 0.891 | 0.941 | 0.439 | 0.561 | 0.510 |
| H12 | 0.378 | 0.622 | 0.572 | 0.092 | 0.858 | 0.908 | 0.414 | 0.586 | 0.536 |
| H13 | 0.445 | 0.555 | 0.504 | 0.059 | 0.891 | 0.941 | 0.888 | 0.112 | 0.060 |
| H14 | 0.800 | 0.200 | 0.150 | 0.200 | 0.750 | 0.800 | 0.852 | 0.148 | 0.093 |
| H15 | 0.800 | 0.200 | 0.150 | 0.072 | 0.878 | 0.928 | 0.865 | 0.135 | 0.081 |
| H16 | 0.800 | 0.200 | 0.150 | 0.092 | 0.858 | 0.908 | 0.822 | 0.178 | 0.125 |
| H17 | 0.414 | 0.586 | 0.536 | 0.059 | 0.891 | 0.941 | 1.000 | 0.000 | 0.000 |
| H18 | 0.650 | 0.350 | 0.300 | 0.081 | 0.869 | 0.919 | 0.869 | 0.131 | 0.077 |
| H19 | 0.414 | 0.586 | 0.536 | 0.200 | 0.750 | 0.800 | 0.200 | 0.750 | 0.800 |
| H20 | 0.650 | 0.350 | 0.300 | 0.050 | 0.900 | 0.950 | 1.000 | 0.000 | 0.000 |
| H21 | 0.650 | 0.350 | 0.300 | 0.072 | 0.878 | 0.928 | 0.650 | 0.350 | 0.300 |
| H22 | 0.650 | 0.350 | 0.300 | 0.092 | 0.858 | 0.908 | 0.852 | 0.148 | 0.093 |
| H23 | 0.378 | 0.622 | 0.572 | 0.092 | 0.858 | 0.908 | 0.650 | 0.350 | 0.300 |
| H24 | 0.800 | 0.200 | 0.150 | 0.079 | 0.871 | 0.921 | 0.888 | 0.112 | 0.060 |
| H25 | 0.650 | 0.350 | 0.300 | 0.059 | 0.891 | 0.941 | 0.852 | 0.148 | 0.093 |
| H26 | 0.650 | 0.350 | 0.300 | 0.092 | 0.858 | 0.908 | 0.650 | 0.350 | 0.300 |
| H27 | 0.650 | 0.350 | 0.300 | 0.081 | 0.869 | 0.919 | 0.865 | 0.135 | 0.081 |
| H28 | 0.900 | 0.100 | 0.050 | 0.439 | 0.561 | 0.510 | 0.200 | 0.750 | 0.800 |
| H29 | 0.650 | 0.350 | 0.300 | 0.081 | 0.869 | 0.919 | 0.865 | 0.135 | 0.081 |
| H30 | 0.650 | 0.350 | 0.300 | 0.059 | 0.891 | 0.941 | 0.822 | 0.178 | 0.125 |
| H31 | 0.650 | 0.350 | 0.300 | 0.079 | 0.871 | 0.921 | 0.865 | 0.135 | 0.081 |

**Table 7.5** Evaluations of risk parameter weights for three OHS experts

| Expert | Probability | Exposure | Consequence |
|---|---|---|---|
| E-1 | U | I | VI |
| E-2 | U | I | I |
| E-3 | M | M | I |
| Weight | (0.378, 0.572, 0.668) | (0.713, 0.286, 0.284) | (0.852, 0.158, 0.129) |

**Table 7.6** Aggregated weighted SVN values of hazards with respect to Fine–Kinney risk parameters

| Hazard | Probability | | | Exposure | | | Consequence | | |
|---|---|---|---|---|---|---|---|---|---|
| H1 | 0.303 | 0.114 | 0.100 | 0.616 | 0.039 | 0.023 | 0.170 | 0.118 | 0.103 |
| H2 | 0.340 | 0.057 | 0.033 | 0.143 | 0.214 | 0.227 | 0.756 | 0.018 | 0.008 |
| H3 | 0.246 | 0.200 | 0.200 | 0.143 | 0.214 | 0.227 | 0.170 | 0.118 | 0.103 |
| H4 | 0.246 | 0.200 | 0.200 | 0.143 | 0.214 | 0.227 | 0.554 | 0.055 | 0.039 |
| H5 | 0.340 | 0.057 | 0.033 | 0.065 | 0.245 | 0.258 | 0.852 | 0.000 | 0.000 |
| H6 | 0.303 | 0.114 | 0.100 | 0.057 | 0.249 | 0.261 | 0.701 | 0.028 | 0.016 |
| H7 | 0.303 | 0.114 | 0.100 | 0.463 | 0.100 | 0.085 | 0.701 | 0.028 | 0.016 |
| H8 | 0.246 | 0.200 | 0.200 | 0.295 | 0.168 | 0.152 | 0.170 | 0.118 | 0.103 |
| H9 | 0.143 | 0.355 | 0.382 | 0.317 | 0.159 | 0.143 | 0.756 | 0.018 | 0.008 |
| H10 | 0.303 | 0.114 | 0.100 | 0.143 | 0.214 | 0.227 | 0.554 | 0.055 | 0.039 |
| H11 | 0.159 | 0.331 | 0.353 | 0.042 | 0.255 | 0.268 | 0.374 | 0.088 | 0.066 |
| H12 | 0.143 | 0.355 | 0.382 | 0.065 | 0.245 | 0.258 | 0.353 | 0.092 | 0.069 |
| H13 | 0.168 | 0.317 | 0.337 | 0.042 | 0.255 | 0.268 | 0.756 | 0.018 | 0.008 |
| H14 | 0.303 | 0.114 | 0.100 | 0.143 | 0.214 | 0.227 | 0.726 | 0.023 | 0.012 |
| H15 | 0.303 | 0.114 | 0.100 | 0.051 | 0.251 | 0.264 | 0.737 | 0.021 | 0.010 |
| H16 | 0.303 | 0.114 | 0.100 | 0.065 | 0.245 | 0.258 | 0.701 | 0.028 | 0.016 |
| H17 | 0.157 | 0.335 | 0.358 | 0.042 | 0.255 | 0.268 | 0.852 | 0.000 | 0.000 |
| H18 | 0.246 | 0.200 | 0.200 | 0.057 | 0.249 | 0.261 | 0.740 | 0.021 | 0.010 |
| H19 | 0.157 | 0.335 | 0.358 | 0.143 | 0.214 | 0.227 | 0.170 | 0.118 | 0.103 |
| H20 | 0.246 | 0.200 | 0.200 | 0.036 | 0.257 | 0.270 | 0.852 | 0.000 | 0.000 |
| H21 | 0.246 | 0.200 | 0.200 | 0.051 | 0.251 | 0.264 | 0.554 | 0.055 | 0.039 |
| H22 | 0.246 | 0.200 | 0.200 | 0.065 | 0.245 | 0.258 | 0.726 | 0.023 | 0.012 |
| H23 | 0.143 | 0.355 | 0.382 | 0.065 | 0.245 | 0.258 | 0.554 | 0.055 | 0.039 |
| H24 | 0.303 | 0.114 | 0.100 | 0.056 | 0.249 | 0.262 | 0.756 | 0.018 | 0.008 |
| H25 | 0.246 | 0.200 | 0.200 | 0.042 | 0.255 | 0.268 | 0.726 | 0.023 | 0.012 |
| H26 | 0.246 | 0.200 | 0.200 | 0.065 | 0.245 | 0.258 | 0.554 | 0.055 | 0.039 |
| H27 | 0.246 | 0.200 | 0.200 | 0.057 | 0.249 | 0.261 | 0.737 | 0.021 | 0.010 |
| H28 | 0.340 | 0.057 | 0.033 | 0.313 | 0.160 | 0.145 | 0.170 | 0.118 | 0.103 |
| H29 | 0.246 | 0.200 | 0.200 | 0.057 | 0.249 | 0.261 | 0.737 | 0.021 | 0.010 |
| H30 | 0.246 | 0.200 | 0.200 | 0.042 | 0.255 | 0.268 | 0.701 | 0.028 | 0.016 |
| H31 | 0.246 | 0.200 | 0.200 | 0.056 | 0.249 | 0.262 | 0.737 | 0.021 | 0.010 |

**Table 7.7** The RNPIS and RNNIS values

| | Probability | Exposure | Consequence |
|---|---|---|---|
| RNPIS | (0.340, 0.057, 0.033) | (0.616, 0.039, 0.023) | (0.852, 0.000, 0.000) |
| RNNIS | (0.143, 0.355, 0.382) | (0.036, 0.257, 0.270) | (0.170, 0.118, 0.103) |

**Table 7.8** Distance measures and closeness coefficient of each hazard

| Hazard | d_Eucl$^+$ | d_Eucl$^-$ | $C_i^*$ | Rank |
|---|---|---|---|---|
| H1 | 0.235 | 0.260 | 0.525 | 6 |
| H2 | 0.185 | 0.264 | 0.588 | 2 |
| H3 | 0.306 | 0.096 | 0.238 | 27 |
| H4 | 0.223 | 0.163 | 0.422 | 22 |
| H5 | 0.211 | 0.287 | 0.576 | 3 |
| H6 | 0.223 | 0.226 | 0.504 | 11 |
| H7 | 0.084 | 0.279 | 0.768 | 1 |
| H8 | 0.276 | 0.132 | 0.324 | 25 |
| H9 | 0.205 | 0.228 | 0.527 | 5 |
| H10 | 0.211 | 0.192 | 0.477 | 15 |
| H11 | 0.314 | 0.071 | 0.185 | 28 |
| H12 | 0.319 | 0.063 | 0.166 | 29 |
| H13 | 0.266 | 0.202 | 0.432 | 20 |
| H14 | 0.189 | 0.237 | 0.555 | 4 |
| H15 | 0.222 | 0.236 | 0.515 | 8 |
| H16 | 0.220 | 0.226 | 0.507 | 10 |
| H17 | 0.269 | 0.233 | 0.464 | 18 |
| H18 | 0.232 | 0.214 | 0.480 | 12 |
| H19 | 0.334 | 0.043 | 0.113 | 30 |
| H20 | 0.236 | 0.249 | 0.513 | 9 |
| H21 | 0.252 | 0.157 | 0.384 | 24 |
| H22 | 0.230 | 0.209 | 0.477 | 16 |
| H23 | 0.288 | 0.132 | 0.314 | 26 |
| H24 | 0.220 | 0.242 | 0.524 | 7 |
| H25 | 0.238 | 0.209 | 0.468 | 17 |
| H26 | 0.248 | 0.158 | 0.389 | 23 |
| H27 | 0.232 | 0.213 | 0.478 | 13 |
| H28 | 0.261 | 0.198 | 0.431 | 21 |
| H29 | 0.232 | 0.213 | 0.478 | 13 |
| H30 | 0.240 | 0.201 | 0.456 | 19 |
| H31 | 0.233 | 0.213 | 0.478 | 14 |

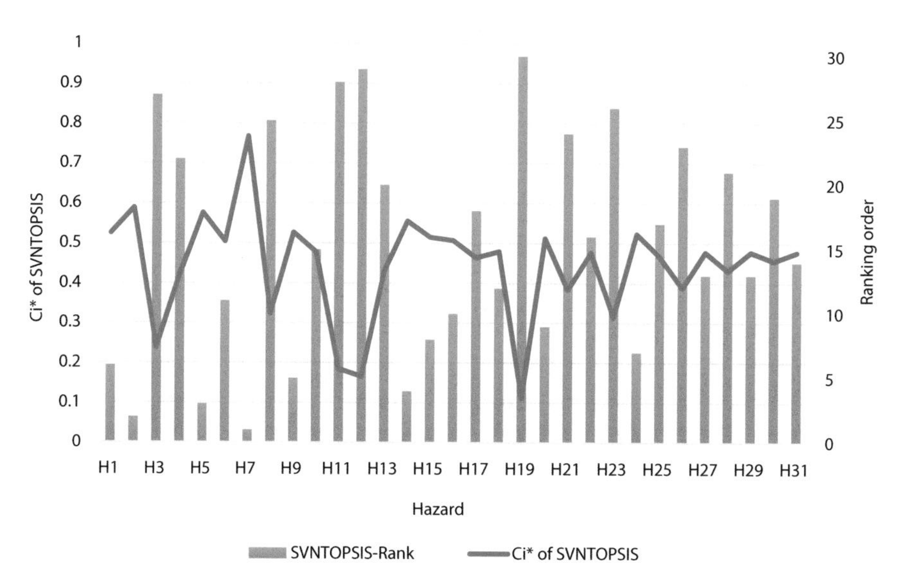


**Fig. 7.2** Final SVNTOPSIS scores and rankings of each hazard

## *7.3.2 Validation Study on the Results*

To test the validity of the proposed approach, a sensitivity analysis is carried on the results. This analysis provides knowledge related to the weights of Fine–Kinney risk parameters. Three additional weight scenarios except the base weight scenario are considered. The cases and their corresponding weight vectors in SVN numbers are provided in Table 7.9.

Base weight scenario shows the original weight values risk parameters in Fine–Kinney achieved from the study. The other scenarios show different changes in weight values. The results for hazards ranking in different scenarios are shown in Table 7.10 and Fig. 7.3.

Figure 7.3 and Table 7.10 demonstrate that in all scenarios except the base weight scenario the most important hazard is H-7. In scenario 1 and scenario 3, as consequence weight is the highest, the ranking of hazards are mostly close. It is also noted

**Table 7.9** The risk parameter weights in Fine–Kinney for the considered scenarios

| Scenario | Weight of risk parameter of Fine–Kinney in SVN numbers | | |
|---|---|---|---|
| | Probability | Exposure | Consequence |
| Base weight scenario | (0.378, 0.572, 0.668) | (0.713, 0.286, 0.284) | (0.852, 0.158, 0.129) |
| Scenario 1 | (0.713, 0.286, 0.284) | (0.378, 0.572, 0.668) | (0.852, 0.158, 0.129) |
| Scenario 2 | (0.378, 0.572, 0.668) | (0.852, 0.158, 0.129) | (0.713, 0.286, 0.284) |
| Scenario 3 | (0.852, 0.158, 0.129) | (0.378, 0.572, 0.668) | (0.713, 0.286, 0.284) |

**Table 7.10** The results for hazards ranking according to three scenarios

| Hazard | Base weight scenario | Scenario 1 | Scenario 2 | Scenario 3 |
|---|---|---|---|---|
| H1 | 6 | 2 | 3 | 2 |
| H2 | 2 | 4 | 2 | 3 |
| H3 | 27 | 27 | 27 | 27 |
| H4 | 22 | 22 | 21 | 22 |
| H5 | 3 | 5 | 4 | 5 |
| H6 | 11 | 11 | 11 | 11 |
| H7 | 1 | 1 | 1 | 1 |
| H8 | 25 | 25 | 25 | 24 |
| H9 | 5 | 3 | 6 | 4 |
| H10 | 15 | 17 | 12 | 13 |
| H11 | 28 | 28 | 28 | 28 |
| H12 | 29 | 29 | 29 | 29 |
| H13 | 20 | 21 | 22 | 21 |
| H14 | 4 | 6 | 5 | 6 |
| H15 | 8 | 9 | 8 | 9 |
| H16 | 10 | 10 | 9 | 10 |
| H17 | 18 | 12 | 19 | 18 |
| H18 | 12 | 13 | 13 | 14 |
| H19 | 30 | 30 | 30 | 30 |
| H20 | 9 | 7 | 10 | 8 |
| H21 | 24 | 24 | 24 | 25 |
| H22 | 16 | 16 | 16 | 17 |
| H23 | 26 | 26 | 26 | 26 |
| H24 | 7 | 8 | 7 | 7 |
| H25 | 17 | 18 | 18 | 19 |
| H26 | 23 | 23 | 23 | 23 |
| H27 | 13 | 14 | 14 | 15 |
| H28 | 21 | 20 | 17 | 12 |
| H29 | 13 | 14 | 14 | 15 |
| H30 | 19 | 19 | 20 | 20 |
| H31 | 14 | 15 | 15 | 16 |

that, according to a correlation analysis, which measures the association between rankings of hazards, there is a strong and significant positive correlation between the scenarios. All Spearman rank correlation coefficient (RHO) values are higher than 0.966. Results of this correlation analysis can be found in Table 7.11.

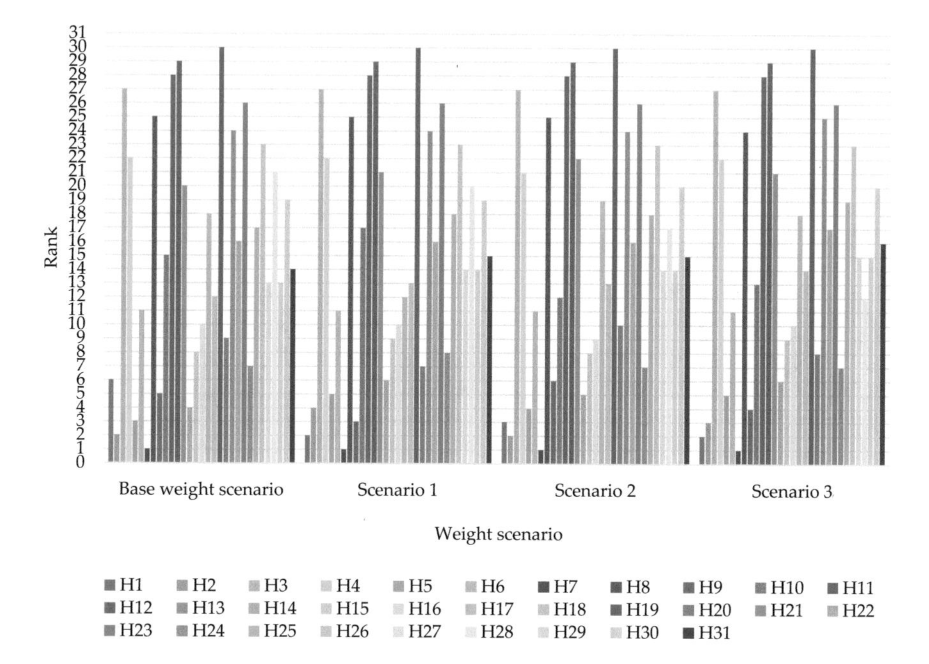


**Fig. 7.3** Results of sensitivity analysis

**Table 7.11** Results of Spearman rank correlation analysis

| Scenario | Base weight scenario | Scenario 1 | Scenario 2 | Scenario 3 |
|---|---|---|---|---|
| Base weight scenario | – | 0.980 | 0.988 | 0.967 |
| Scenario 1 | | – | 0.972 | 0.966 |
| Scenario 2 | | | – | 0.988 |
| Scenario 3 | | | | – |

## 7.4 Python Implementation of the Proposed Approach

```
# Chapter 7
# import required libraries
import numpy as np


    n_criteria = 3  # P, E, C
    ns_value = 3  # SVN
    n_expert = 3  # E-1,E-2, E-3
    n_hazard = 31
```

```
# linguistic relation terms
linguistic_rt = [ # SVN number
  ["VI", 0.90, 0.10, 0.10],
  ["I", 0.80, 0.20, 0.15],
  ["M", 0.50, 0.40, 0.45],
  ["U", 0.35, 0.60, 0.70],
  ["VU", 0.10, 0.80, 0.90]
]

e_weight = [ # experts' weight
  ["E-1", "VI", 0, 0],
  ["E-2", "I", 0, 0],
  ["E-3", "U", 0, 0]
]

# elm: expert linguistic matrix
elm = [ # P  E  C
  ["expert1", "U", "I", "VI"],
  ["expert2", "U", "M", "I"],
  ["expert3", "M", "I", "I"]
]

linguistic_terms = [ # SVN number
  ["EH", 1.00, 0.00, 0.00],
  ["VH", 0.90, 0.10, 0.05],
  ["H", 0.80, 0.20, 0.15],
  ["MH", 0.65, 0.35, 0.30],
  ["M", 0.50, 0.50, 0.45],
  ["ML", 0.35, 0.65, 0.60],
  ["L", 0.20, 0.75, 0.80],
  ["VL", 0.10, 0.85, 0.90],
  ["EL", 0.05, 0.90, 0.95]
]
```

```
# ehe: 1st expert's hazard evaluation
ehe1 = [ # Hazard id, P, E, C,
    ["H1", "H", "H", "L"],
    ["H2", "VH", "L", "VH"],
    ["H3", "MH", "L", "L"],
    ["H4", "MH", "L", "MH"],
    ["H5", "VH", "VL", "EH"],
    ["H6", "H", "VL", "H"],
    ["H7", "H", "MH", "H"],
    ["H8", "MH", "ML", "L"],
    ["H9", "ML", "M", "VH"],
    ["H10", "H", "L", "MH"],
    ["H11", "M", "EL", "ML"],
    ["H12", "ML", "VL", "ML"],
    ["H13", "M", "EL", "VH"],
    ["H14", "H", "L", "VH"],
    ["H15", "H", "VL", "H"],
    ["H16", "H", "VL", "H"],
    ["H17", "ML", "EL", "EH"],
    ["H18", "MH", "VL", "VH"],
    ["H19", "ML", "L", "L"],
    ["H20", "MH", "EL", "EH"],
    ["H21", "MH", "VL", "MH"],
    ["H22", "MH", "VL", "VH"],
    ["H23", "ML", "VL", "MH"],
    ["H24", "H", "EL", "VH"],
    ["H25", "MH", "EL", "VH"],
    ["H26", "MH", "VL", "MH"],
    ["H27", "MH", "VL", "H"],
    ["H28", "VH", "ML", "L"],
    ["H29", "MH", "VL", "H"],
    ["H30", "MH", "EL", "H"],
    ["H31", "MH", "EL", "H"]
]
```

```
# ehe: 2nd expert's hazard evaluation
ehe2 = [ # Hazard id,P, E, C
  ["H1", "H", "VH", "L"],
  ["H2", "VH", "L", "VH"],
  ["H3", "MH", "L", "L"],
  ["H4", "MH", "L", "MH"],
  ["H5", "VH", "VL", "EH"],
  ["H6", "H", "EL", "H"],
  ["H7", "H", "MH", "H"],
  ["H8", "MH", "M", "L"],
  ["H9", "ML", "ML", "VH"],
  ["H10", "H", "L", "MH"],
  ["H11", "ML", "EL", "M"],
  ["H12", "ML", "VL", "M"],
  ["H13", "ML", "EL", "VH"],
  ["H14", "H", "L", "H"],
  ["H15", "H", "EL", "VH"],
  ["H16", "H", "VL", "H"],
  ["H17", "M", "EL", "EH"],
  ["H18", "MH", "EL", "H"],
  ["H19", "M", "L", "L"],
  ["H20", "MH", "EL", "EH"],
  ["H21", "MH", "EL", "MH"],
  ["H22", "MH", "VL", "H"],
  ["H23", "ML", "VL", "MH"],
  ["H24", "H", "VL", "VH"],
  ["H25", "MH", "EL", "H"],
  ["H26", "MH", "VL", "MH"],
  ["H27", "MH", "EL", "VH"],
  ["H28", "VH", "M", "L"],
  ["H29", "MH", "EL", "VH"],
  ["H30", "MH", "EL", "H"],
  ["H31", "MH", "VL", "VH"]
]
```

```
# ehe: 3rd expert's hazard evaluation
ehe3 = [ # Hazard id, P, E, C
    ["H1", "H", "VH", "L"],
    ["H2", "VH", "L", "H"],
    ["H3", "MH", "L", "L"],
    ["H4", "MH", "L", "MH"],
    ["H5", "VH", "EL", "EH"],
    ["H6", "H", "VL", "VH"],
    ["H7", "H", "MH", "VH"],
    ["H8", "MH", "ML", "L"],
    ["H9", "M", "M", "H"],
    ["H10", "H", "L", "MH"],
    ["H11", "ML", "VL", "M"],
    ["H12", "M", "EL", "ML"],
    ["H13", "M", "VL", "H"],
    ["H14", "H", "L", "H"],
    ["H15", "H", "EL", "VH"],
    ["H16", "H", "EL", "VH"],
    ["H17", "ML", "VL", "EH"],
    ["H18", "MH", "VL", "VH"],
    ["H19", "ML", "L", "L"],
    ["H20", "MH", "EL", "EH"],
    ["H21", "MH", "EL", "MH"],
    ["H22", "MH", "EL", "H"],
    ["H23", "M", "EL", "MH"],
    ["H24", "H", "VL", "H"],
    ["H25", "MH", "VL", "H"],
    ["H26", "MH", "EL", "MH"],
    ["H27", "MH", "VL", "VH"],
    ["H28", "VH", "M", "L"],
    ["H29", "MH", "VL", "VH"],
    ["H30", "MH", "VL", "VH"],
    ["H31", "MH", "VL", "VH"]
]


def rank(vector, da):  # da -1:descending, 1:ascending
    order = np.zeros([len(vector), 1])
    unique_val = da * np.sort(da * np.unique(vector))
    for ix in range(0, len(unique_val)):
        order[np.argwhere(vector == unique_val[ix])] = ix +1
    return order


def print_result(order, vector):
    print('Hazard Id, Rank, Value')
    for ix in range(0, len(order)):
        print(ehe1[ix][0], ', ', int(order[ix]), ', ', vector[ix])
```

```
sm = 0
for j in range(0, n_expert):
  for lrt in linguistic_rt:
    if e_weight[j][1] == lrt[0]:
      e_weight[j][2] = np.sqrt((np.square(1 - lrt[1])
                     + np.square(lrt[2])
                     + np.square(lrt[3])) /3)
      sm += e_weight[j][2]


sm = 3 - sm
for j in range(0, n_expert):
  e_weight[j][3] = (1 - e_weight[j][2]) / sm # expert weights


# enm: expert numeric matrix
enm = np.zeros([n_expert, n_criteria, ns_value], dtype=float)
for i in range(0, n_expert):
  for j in range(0, n_criteria):
    for lt in linguistic_rt:
      if elm[i][j + 1] == lt[0]:
        enm[i][j] = lt[1:]


criteria_matrix = np.ones([n_criteria, ns_value], dtype=float)
for i in range(0, n_expert):
  criteria_matrix *= np.power(1 - enm[i], e_weight[i][3])
criteria_matrix = 1 - criteria_matrix


ehe = np.asarray([ehe1, ehe2, ehe3])
ehe_numeric = np.zeros([n_expert, n_hazard, n_criteria, ns_value], dtype=float)
for i in range(0, n_expert):
  for j in range(0, n_hazard):
    for k in range(0, n_criteria):
      for lt in linguistic_terms:
        if ehe[i][j][k + 1] == lt[0]:
          ehe_numeric[i][j][k] = lt[1:]


ehe_numeric[:, :, :, 0] = 1 - ehe_numeric[:, :, :, 0]
aggregated_matrix = np.ones([n_hazard, n_criteria, ns_value], dtype=float)
for i in range(0, n_expert):
  aggregated_matrix *= np.power(ehe_numeric[i], e_weight[i][3])
aggregated_matrix[:, :, 0] = 1 - aggregated_matrix[:, :, 0]


weighted_matrix = np.zeros([n_hazard, n_criteria, ns_value], dtype=float)
for i in range(0, n_hazard):
  for j in range(0, n_criteria):
    for k in range(0, ns_value):
      weighted_matrix[i][j][k] = aggregated_matrix[i][j][k] \
                  * criteria_matrix[j][k]
```

```
NRPIS = []
NRNIS = []
for i in range(0, n_criteria):
  for j in range(0, ns_value):
    max_val = np.max(weighted_matrix[:, i, j])
    min_val = np.min(weighted_matrix[:, i, j])
    if j ==0:
      NRPIS.append(max_val)
      NRNIS.append(min_val)
    else:
      NRPIS.append(min_val)
      NRNIS.append(max_val)


d_plus = []
min_rank = [[0 for j in range(0, n_criteria)]
      for i in range(0, n_hazard)]


temp_val = np.reshape(weighted_matrix, [n_hazard, n_criteria * ns_value])
d_plus = np.sqrt((1 / 6 * n_criteria)) * np.sum(np.square(temp_val - NRPIS), axis=1))
d_minus = np.sqrt((1 / 6 * n_criteria)) * np.sum(np.square(temp_val - NRNIS), axis=1))


ci = d_minus / (d_plus + d_minus)


hazard_rank = rank(ci, -1)
print_result(hazard_rank, ci)
```

```
'''
Output:
Hazard Id, Rank, Value
H1 , 6 , 0.5249819676876027
H2 , 2 , 0.5883799841930879
H3 , 27 , 0.23838786372496265
H4 , 22 , 0.4215299924258605
H5 , 3 , 0.575749531112042
H6 , 11 , 0.5037733157023486
H7 , 1 , 0.7676378346967132
H8 , 25 , 0.3239310391219617
H9 , 5 , 0.5269643293672033
H10 , 15 , 0.4771729844442515
H11 , 28 , 0.1847455069598435
H12 , 29 , 0.16610788508942584
H13 , 20 , 0.43173372303090635
H14 , 4 , 0.5554399889841435
H15 , 8 , 0.5149821834711917
H16 , 10 , 0.5072296101229544
H17 , 18 , 0.4643576136829114
H18 , 12 , 0.4796491813986901
H19 , 30 , 0.11298872115932172
H20 , 9 , 0.5126485003250184
H21 , 24 , 0.38422886733395445
H22 , 16 , 0.47682740591057543
H23 , 26 , 0.3142772205794686
H24 , 7 , 0.5239267411160863
H25 , 17 , 0.46760396985626024
H26 , 23 , 0.38901484122226593
H27 , 13 , 0.4782661963294243
H28 , 21 , 0.43124940275656015
H29 , 13 , 0.4782661963294243
H30 , 19 , 0.45614718526497283
H31 , 14 , 0.47768734548657354
'''
```

## References


1. Smarandache, F. (2002). Neutrosophy and neutrosophic logic. In *First international conference on neutrosophy, neutrosophic logic, set, probability, and statistics* (pp. 338–353). Gallup, NM: University of New Mexico.
2. Biswas, P., Pramanik, S., & Giri, B. C. (2016). TOPSIS method for multi-attribute group decision-making under single-valued neutrosophic environment. *Neural Computing and Applications, 27,* 727–737.
3. Liu, P., & Wang, Y. (2014). Multiple attribute decision-making method based on single valued neutrosophic normalized weighted Bonferroni mean. *Neural Computing and Applications, 25,* 2001–2010.
4. Majumdar, P., & Samanta, S. K. (2014). On similarity and entropy of neutrosophic sets. *Journal of Intelligent & Fuzzy Systems, 26,* 1245–1252.

5. Yoon, K., & Hwang, C. (1981). *TOPSIS (technique for order preference by similarity to ideal solution)–a multiple attribute decision making, w: Multiple attribute decision making–methods and applications, a state-of-the-at survey*. Berlin: Springer.
6. Yong, D. (2006). Plant location selection based on fuzzy TOPSIS. *The International Journal of Advanced Manufacturing Technology, 28,* 839–844.
7. Chen, T.-Y., & Tsao, C.-Y. (2008). The interval-valued fuzzy TOPSIS method and experimental analysis. *Fuzzy Sets and Systems, 159,* 1410–1428.
8. Chen, S.-M., & Lee, L.-W. (2010). Fuzzy multiple attributes group decision-making based on the interval type-2 TOPSIS method. *Expert Systems with Applications, 37,* 2790–2798.
9. Celik, E., Bilisik, O. N., Erdogan, M., et al. (2013). An integrated novel interval type-2 fuzzy MCDM method to improve customer satisfaction in public transportation for Istanbul. *Transportation Research Part E: Logistics and Transportation Review, 58,* 28–51.
10. Cevik Onar, S., Oztaysi, B., & Kahraman, C. (2014). Strategic decision selection using hesitant fuzzy TOPSIS and interval type-2 fuzzy AHP: A case study. *International Journal of Computational intelligence systems, 7,* 1002–1021.
11. Boran, F., Boran, K., & Menlik, T. (2012). The evaluation of renewable energy technologies for electricity generation in Turkey using intuitionistic fuzzy TOPSIS. *Energy Sources, Part B: Economics, Planning and Policy, 7,* 81–90.
12. Ozdemir, Y., Gul, M., & Celik, E. (2017). Assessment of occupational hazards and associated risks in fuzzy environment: A case study of a university chemical laboratory. *Human and Ecological Risk Assessment: An International Journal, 23,* 895–924.
13. Ak, M. F., & Gul, M. (2018). AHP–TOPSIS integration extended with Pythagorean fuzzy sets for information security risk analysis. *Complex & Intelligent Systems* 1–14.
14. Gul, M., & Ak, M. F. (2018). A comparative outline for quantifying risk ratings in occupational health and safety risk assessment. *Journal of Cleaner Production, 196,* 653–664.
15. Oz, N. E., Mete, S., Serin, F., & Gul, M. (2018). Risk assessment for clearing and grading process of a natural gas pipeline project: An extended TOPSIS model with Pythagorean fuzzy sets for prioritizing hazards. *Human and Ecological Risk Assessment: An International Journal* 1–18.
16. Mete, S. (2019). Assessing occupational risks in pipeline construction using FMEA-based AHP-MOORA integrated approach under Pythagorean fuzzy environment. *Human and Ecological Risk Assessment: An International Journal* 1–16. https://doi.org/10.1080/10807039.2018.1546115.
17. Mete, S., Serin, F., Oz, N. E., & Gul, M. (2019). A decision-support system based on Pythagorean fuzzy VIKOR for occupational risk assessment of a natural gas pipeline construction. *Journal of Natural Gas Science and Engineering, 71,* 102979.
18. Gul, M., Ak, M. F., & Guneri, A. F. (2019). Pythagorean fuzzy VIKOR-based approach for safety risk assessment in mine industry. *Journal of Safety Research, 69,* 135–153.
19. Gul, M., Guneri, A. F., & Baskan, M. (2018). An occupational risk assessment approach for construction and operation period of wind turbines. *Global Journal of Environmental Science and Management, 4*(3), 281–298.

# Chapter 8
# Fine–Kinney-Based Occupational Risk Assessment Using Interval Type-2 Fuzzy QUALIFLEX

**Abstract** In this chapter, we improved Fine–Kinney occupational risk assessment approach with interval type-2 fuzzy QUALIFLEX (IT2FQUALIFLEX). QUALIFLEX is an outranking multi-attribute decision-making method proposed by an extension of the Paelinck's (Pap Reg Sci Assoc 36:59–74,[1]), generalized Jacquet-Lagreze's permutation method. Similar to other outranking solution-based approaches, it considers the solution which is a comparison of hazards. In this chapter, we adapted the interval type-2 fuzzy sets (IT2FSs) into QUALIFLEX as it reflects the uncertainty well in decision-making. IT2FQUALIFLEX algorithm under the Fine–Kinney concept provides a useful and solid approach to the occupational health and safety risk assessment. In addition to proposing this new approach, a case study is performed in the chrome plating unit. A validation is also performed in this study. Finally, the proposed approach is implemented in Python.

## 8.1 Interval Type-2 Fuzzy QUALIFLEX (IT2FQUALIFLEX)

Prior to defining the proposed IT2FQUALIFLEX, a general view on the classic QUALIFLEX method is provided.

### *8.1.1 QUALIFLEX Method*

The classical QUALIFLEX was initially proposed as an outranking approach. It is an extension of the Paelinck's [1], generalized Jacquet-Lagreze's permutation method. There is $n!$ permutations are available for $n$ hazards which test each possible ranking of the hazards.

The main steps can be summarized in the following steps [2]:

1. Define the hazards and parameters for constructing the decision matrix,
2. Generate all of the possible permutations of the hazards,



M. Gul et al., *Fine–Kinney-Based Fuzzy Multi-criteria Occupational Risk Assessment*, Studies in Fuzziness and Soft Computing 398,
https://doi.org/10.1007/978-3-030-52148-6_8

3. Calculate the concordance/discordance index for each pair of hazards in each permutation,
4. Calculate the weighted the concordance/discordance index using the importance weight of parameters.
5. Compute the final ranking and select the highest value permutation.

### 8.1.2 *Interval Type-2 Fuzzy QUALIFLEX*

Different versions of the QUALIFLEX are proposed for different application areas. Chen et al. [3] and Wang et al. [8] proposed QUALIFLEX under IT2FSs for medical decision-making problem. Interval-valued Pythagorean fuzzy QUALIFLEX is proposed for risk evaluation [9]. Interval-valued intuitionistic fuzzy QUALIFLEX method is proposed for the selection of the most suitable bridge construction method [10] and the selection of a landfill site [14]. Li and Wang [11], Zhang and Xu [12] Zhang et al. [13] and Zhang [19] proposed QUALIFLEX under hesitant fuzzy environment for selecting green suppliers. Demirel et al. [15] and Gumus et al. [16] applied QUALIFLEX under IT2FSs for evaluating the performance of ballast water treatment and passenger satisfaction evaluation for rail transit lines, respectively. Chen et al. [3] used the signed distance in IT2FQUALIFLEX approach. In this paper, we applied a ranking value method [4, 5], and used their linguistic terms and corresponding IT2FSs for QUALIFLEX approach. The fuzziness in type-2 fuzzy numbers gives more reasonable consequences comparing to type-1 fuzzy numbers in terms of demonstrating uncertainties in the practical applications [17, 18]. The proposed approach based on interval type-2 fuzzy numbers (IT2FNs) is given as follows [3]:

**Step 1**: The hazards ($H = \{H_1, H_2, \ldots, H_n\}$) and parameters ($C = \{c_1, c_2, \ldots, c_m\}$) are determined for evaluating the risk assessment process.

**Step 2**: The linguistic variables are converted to the rating of the hazards with respect to each parameter.

**Step 3**: The linguistic evaluation of the hazards to each parameter is implemented by OHS experts.

**Step 4**: The linguistic variables are converted to IT2FNs to obtain the rating $H_{ij}$ of the hazards $H_j$ on the parameters $c_i$. $\tilde{\tilde{H}}_{ij}$ is an IT2FNs, $i = 1, 2, \ldots, m, p = 1, 2, \ldots, k$, k symbolizes the number of OHS, and $\tilde{\tilde{H}}_{ij}$ is expressed as follows:

$$\begin{aligned}\tilde{\tilde{H}}_{ij} &= \left[\tilde{H}_{ij}^U, \tilde{H}_{ij}^L\right] \\ &= \left[\left(h_{1ij}^U, h_{2ij}^U, h_{3ij}^U, h_{4ij}^U; R_1\left(\tilde{\tilde{H}}_1^U\right), R_2\left(\tilde{\tilde{H}}_1^U\right)\right), \left(h_{1ij}^L, h_{2ij}^L, h_{3ij}^L, h_{4ij}^L; R_1\left(\tilde{H}_1^L\right), R_2\left(\tilde{H}_1^L\right)\right)\right]\end{aligned} \quad (8.1)$$

where$1 \leq i \leq m$, $1 \leq j \leq n$. The decision matrix $E_c$ is constructed as follows:

$$E_c = (\tilde{\tilde{H}}_{ij}^{p})_{m\times n} = \begin{array}{c} \\ c_1 \\ c_2 \\ \vdots \\ c_m \end{array}\begin{array}{c} \begin{array}{cccc} H_1 & H_2 & \cdots & H_n \end{array} \\ \begin{bmatrix} \tilde{\tilde{H}}_{11}^{p} & \tilde{\tilde{H}}_{12}^{p} & \cdots & \tilde{\tilde{H}}_{1n}^{p} \\ \tilde{\tilde{H}}_{21}^{p} & \tilde{\tilde{H}}_{22}^{p} & \cdots & \tilde{\tilde{H}}_{2n}^{p} \\ \vdots & \vdots & \vdots & \vdots \\ \tilde{\tilde{H}}_{m1}^{p} & \tilde{\tilde{H}}_{m2}^{p} & \cdots & \tilde{\tilde{H}}_{mn}^{p} \end{bmatrix} \end{array} \quad (8.2)$$

where $\tilde{\tilde{H}}_{ij} = \left(\frac{\tilde{\tilde{H}}_{ij}^{1}+\tilde{\tilde{H}}_{ij}^{2}+\cdots+\tilde{\tilde{H}}_{ij}^{k}}{k}\right)$, $1 \le p \le k$, k presents OHS experts who involve in risk assessment. The characteristics of the hazards $H_{ij}$ are presented by the IT2FNs as follows:

$$H_{ij} = \left\{\left\langle c_i, \left[H_{ij}^{U}, H_{ij}^{L}\right]\right\rangle | c_i \in C, i = 1, 2, \ldots, m\right\}, j = 1, 2, \ldots, n.$$

**Step 5**: The ranking value $Rank\left(\tilde{\tilde{H}}_{ij}\right)$ for hazards $H_{ij}$ in $E_c$ is computed. The ranking of fuzzy numbers has importance on decision-making process. The ranking value $Rank\left(\tilde{\tilde{H}}_{ij}\right)$ of the IT2FSs $\tilde{\tilde{H}}_{ij}$ is specified [4, 5], as follows:

$$\begin{aligned} Rank(\tilde{\tilde{H}}_{ij}) = & M_1\left(\tilde{H}_{ij}^{U}\right) + M_1\left(\tilde{H}_{ij}^{L}\right) + M_2\left(\tilde{H}_{ij}^{U}\right) + M_2\left(\tilde{H}_{ij}^{L}\right) + M_3\left(\tilde{H}_{ij}^{U}\right) \\ & + M_3\left(\tilde{H}_{ij}^{L}\right) - \frac{1}{4}\left(\begin{array}{l} S_1\left(\tilde{H}_{ij}^{U}\right) + S_1\left(\tilde{H}_{ij}^{L}\right) + S_2\left(\tilde{H}_{ij}^{U}\right) + S_2\left(\tilde{H}_{ij}^{L}\right) + S_3\left(\tilde{H}_{ij}^{U}\right) \\ + S_3\left(\tilde{H}_{ij}^{L}\right) + S_4\left(\tilde{H}_{ij}^{U}\right) + S_4\left(\tilde{H}_{ij}^{L}\right) \end{array}\right) \\ & + R_1\left(\tilde{H}_{ij}^{U}\right) + R_1\left(\tilde{H}_{ij}^{L}\right) + R_2\left(\tilde{H}_{ij}^{U}\right) + R_2\tilde{H}\left(_{ij}^{L}\right) \end{aligned} \quad (8.3)$$

where $M_p(\tilde{\tilde{H}}_i^*)$ symbolizes the mean of the elements $h_{ip}^*$ and $h_{i(p+1)}^*$,$M_p(\tilde{\tilde{H}}_i^*) = \left(h_{ip}^* + h_{i(p+1)}^*\right)/2$, $1 \le p \le 3$, symbolizes the standard deviation of the elements $h_{iq}^*$ and $r_{i(q+1)}^*$,$S_q(\tilde{H}_i^*) = \sqrt{\frac{1}{2}\sum_{k=q}^{q+1}\left(h_{ik}^* - \frac{1}{2}\sum_{k=q}^{q+1} h_{ik}^*\right)^2}$, $1 \le q \le 3$,$S_4(\tilde{H}_i^*)$ symbolizes the standard deviation of the elements $h_{i1}^*, h_{i2}^*, h_{i3}^*, h_{i4}^*$, $S_4(\tilde{H}_i^*) = \sqrt{\frac{1}{4}\sum_{k=1}^{4}\left(h_{ik}^* - \frac{1}{4}\sum_{k=1}^{4} h_{ik}^*\right)^2}$ $R_p(\tilde{H}_i^*)$ symbolizes the membership value of the element $h_{i(p+1)}^*$ in the membership function $\tilde{H}_i^*$, $1 \le p \le 3$, $* \in \{U, L\}$, and $1 \le i \le n$.

**Step 6**: The all possible *n!* permutations of the n hazards must be tested. The n$th$ permutation $P_g = (1, 2, \ldots, n!)$ is calculated as follows. The ranking value approach is used to determine the concordance/discordance index. When the hazard $H_{\alpha i}$ is ranked higher than or equal to $H_{\beta i}$, then the evaluation values of $H_{\alpha i}$ and $H_{\beta i}$ for each criterion $c_i \in C$ is determined as follows:

$$\tilde{\tilde{H}}_{\alpha i} = \left[\tilde{H}^U_{\alpha i}, \tilde{H}^L_{\alpha i}\right] = \left[\begin{array}{c}\left(h^U_{1\alpha i}, h^U_{2\alpha i}, h^U_{3\alpha i}, h^U_{4\alpha i}; R_1\left(\tilde{H}^U_{\alpha i}\right), R_2\left(\tilde{H}^U_{\alpha i}\right)\right), \\ \left(h^L_{1\alpha i}, h^L_{2\alpha i}, h^L_{3\alpha i}, h^L_{4\alpha i}; R_1\left(\tilde{H}^L_{\alpha i}\right), R_2\left(\tilde{H}^L_{\alpha i}\right)\right)\end{array}\right]$$ and

$$\tilde{\tilde{H}}_{\beta i} = \left[\tilde{H}^U_{\beta i}, \tilde{H}^L_{\beta i}\right] = \left[\begin{array}{c}\left(h^U_{1\beta i}, h^U_{2\beta i}, h^U_{3\beta i}, h^U_{4\beta i}; R_1\left(\tilde{H}^U_{\beta i}\right), R_2\left(\tilde{H}^U_{\beta i}\right)\right), \\ \left(h^L_{1\beta i}, h^L_{2\beta i}, h^L_{3\beta i}, h^L_{4\beta i}; R_1\left(\tilde{H}^L_{\beta i}\right), R_2\left(\tilde{H}^L_{\beta i}\right)\right)\end{array}\right]$$ The ranking value of $Rank\left(\tilde{\tilde{H}}_{\alpha i}\right)$ and $Rank\left(\tilde{\tilde{H}}_{\beta i}\right)$ is considered to rank $H_{\alpha i}$ and $H_{\beta i}$. There are three different conditions as concordance, ex aequo and discordance with the preorders. If $Rank\left(\tilde{\tilde{H}}_{\alpha i}\right) > Rank\left(\tilde{\tilde{H}}_{\beta i}\right)$, hazard $H_{\alpha i}$ have higher rank than hazard $H_{\beta i}$. On the other hand, if $Rank\left(\tilde{\tilde{H}}_{\alpha i}\right) = Rank\left(\tilde{\tilde{H}}_{\beta i}\right)$, two hazards have the same rank. The other observation is discordance $Rank\left(\tilde{\tilde{H}}_{\alpha i}\right) < Rank\left(\tilde{\tilde{H}}_{\beta i}\right)$ which hazard $H_{\alpha i}$ rank below the hazard $H_{\beta i}$.

**Step 7**: The concordance or discordance index $I^g_i\left(H_\alpha, H_\beta\right)$ is calculated for each pair of $\left(H_\alpha, H_\beta\right)$ in the permutation $P_g$ according to the criterion $c_i \in C$ using Eq. (8.4).

$$I^g_i\left(H_\alpha, H_\beta\right) = Rank(H_\alpha) - Rank\left(H_\beta\right) \tag{8.4}$$

Also, the concordance/discordance index $I^g_i$ for criterion $c_i \in C$ is calculated as follows:

$$I^g_i = \sum_{H_\alpha, H_\beta \in H} I^g_i\left(H_\alpha, H_\beta\right) = \sum_{H_\alpha, H_\beta \in H} \left(Rank\left(\tilde{\tilde{H}}_\alpha\right) - Rank\left(\tilde{\tilde{H}}_\beta\right)\right) \tag{8.5}$$

**Step 8**: The weight of each parameter is obtained by OHS experts. Then, the weighted concordance/discordance index $I^g_i\left(H_\alpha, H_\beta\right) \cdot w_i$ is calculated for each pair of hazards in $n$th permutation.

$$\begin{aligned} I^g\left(H_\alpha, H_\beta\right) &= \sum_{i=1}^{m} I^g_i\left(H_\alpha, H_\beta\right) \cdot w_i \\ &= \sum_{i=1}^{m} I^g_i\left(Rank\left(\tilde{\tilde{H}}_\alpha\right) - Rank\left(\tilde{\tilde{H}}_\beta\right)\right) \cdot w_i \end{aligned} \tag{8.6}$$

**Step 9**: The detailed concordance/discordance index $I^g$ is computed using Eq. (8.7).

$$I^g = \sum_{H_\alpha, H_\beta \in H} \sum_{i=1}^{m} I^g_i\left(H_\alpha, H_\beta\right) \cdot w_i$$

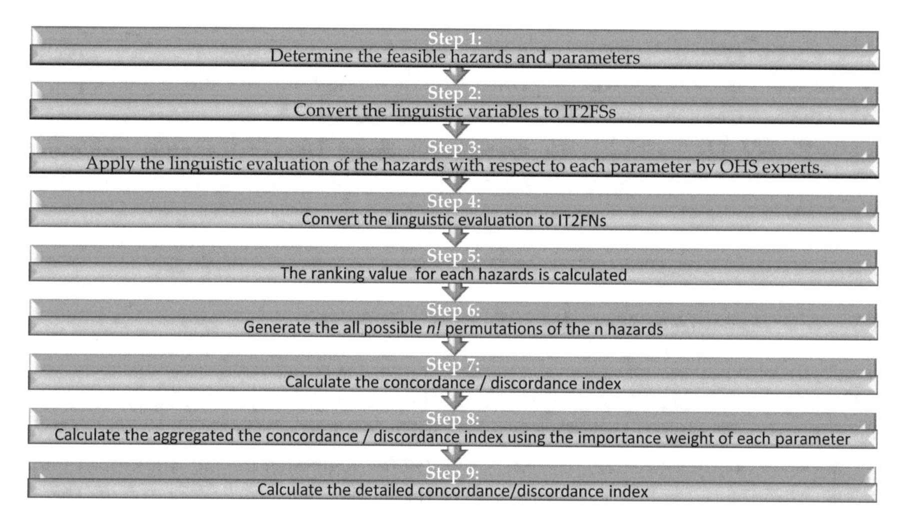


**Fig. 8.1** The main steps of the proposed approach

$$= \sum_{H_\alpha, H_\beta \in H} \sum_{i=1}^{m} I_i^g \left( Rank\left(\tilde{\tilde{H}}_\alpha\right) - Rank\left(\tilde{\tilde{H}}_\beta\right)\right) \cdot w_i \tag{8.7}$$

Then, the detailed concordance and discordance index for all permutations is compared using the calculation of ranking value (using Eq. 8.3)

$$\begin{aligned} Rank(I^g) &= Rank\left( \sum_{H_\alpha, H_\beta \in H} \sum_{i=1}^{m} I_i^g (H_\alpha, H_\beta) \cdot w_i \right) \\ &= Rank\left( \sum_{H_\alpha, H_\beta \in H} \sum_{i=1}^{m} I_i^g \left(Rank(H_{\alpha i}) - Rank(H_{\beta i})\right) \cdot w_i \right) \\ &\text{for } g = 1, 2, \ldots, n!. \end{aligned} \tag{8.8}$$

the optimal ranking order of the hazards for risk assessment is determined using the maximum $Rank(I^g)$. We presented the main steps of the proposed approach in Fig. 8.1.

## 8.2 Case Study

In applying the proposed approach, we demonstrate a case study for the chrome plating unit. The case study is previously studied in [6]. The current study is differentiated from [6], by the aspects of covering the proposed methodology. In [6], Fuzzy

AHP and Fuzzy VIKOR under triangular fuzzy numbers is proposed. Therefore, the existed work is indeed novel in terms of methodology and not the same as [6], in terms of case study demonstration scope and boundaries. As the main method of the approach, IT2FQUALIFLEX was used to rank hazards and their associated risks. In the following, step by step application of the proposed approach to the problem is provided.

### *8.2.1 Application Results*

In Step 1, we have six hazards H = {H−1, H−2, H−3, H−4, H−5 and H−6} and 3 different Fine–Kinney parameters as probability ($P$), exposure ($E$), and consequence ($C$) are used in assessing hazards. Also, the risk list regarding the camouflage coating unit is provided in Table 8.1. The weights of these parameters are derived from [6], as $w_P = 0.289$, $w_E = 0.293$, $w_C = 0.418$. In Step 2, the six-point linguistic variables to evaluate risk assessment of hazards is selected for conversion into IT2FNs from Table 8.2. This scale is obtained from Celik et al. [7]. While using the terms from Table 8.2, the evaluations given in Table 8.3, are obtained to construct the fuzzy aggregated decision matrix. In Step 3, the six hazards were evaluated based on the three parameters using the linguistic variables. In Step 4, linguistic variables were converted into IT2FNs. Then, the decision matrix $E_c$ is obtained. The average IT2F values for the hazards are presented in Table 8.4.

**Table 8.1** The hazard identification list in chrome plating unit

| ID | Hazard identification |
|---|---|
| Hazard-1 | Insufficient ventilation |
| Hazard-2 | Electricity |
| Hazard-3 | Brightness, excessive lighting, insufficient lighting |
| Hazard-4 | Flammable, flammable and explosive environments |
| Hazard-5 | Chemical Liquid + Gas (Alcohol, welding, soldering gas etc.) exposure |
| Hazard-6 | Professional competence/experience |

**Table 8.2** The scale used in assessing hazards

| Linguistic term | Interval type-2 fuzzy numbers |
|---|---|
| Poor (P) | ((0, 1, 1, 3; 1, 1), (0.5, 1, 1, 2; 0.9, 0.9)) |
| Medium Poor (MP) | ((1, 3, 3, 5; 1, 1), (2, 3, 3, 4; 0.9, 0.9)) |
| Medium (M) | ((3, 5, 5, 7; 1, 1), (4, 5, 5, 6; 0.9, 0.9)) |
| Medium good (MG) | ((5, 7, 7, 9; 1, 1), (6, 7, 7, 8; 0.9, 0.9)) |
| Good (G) | ((7, 9, 9, 10; 1, 1), (8, 9, 9, 9.5; 0.9, 0.9)) |
| Very good (VG) | ((9, 10, 10, 10; 1, 1), (9.5, 10, 10, 10; 0.9, 0.9)) |

**Table 8.3** The linguistic initial decision matrix from the OHS experts' consensus

| Hazard ID | Hazard-1 | Hazard-2 | Hazard-3 | Hazard-4 | Hazard-5 | Hazard-6 |
|---|---|---|---|---|---|---|
| Probability | MG | MG | M | G | MG | M |
| Exposure | VG | MP | M | G | G | MP |
| Consequence | M | G | MG | G | MG | MG |

**Table 8.4** Average IT2F values

| Hazard ID | Hazard-1 | Hazard-2 |
|---|---|---|
| Probability | ((5, 7, 7, 9; 1, 1), (6, 7, 7, 8; 0.9, 0.9)) | ((5, 7, 7, 9; 1, 1), (6, 7, 7, 8; 0.9, 0.9)) |
| Exposure | ((9, 10, 10, 10; 1, 1), (9.5, 10, 10, 10; 0.9, 0.9)) | ((1, 3, 3, 5; 1, 1), (2, 3, 3, 4; 0.9, 0.9)) |
| Consequence | ((3, 5, 5, 7; 1, 1), (4, 5, 5, 6; 0.9, 0.9)) | ((7, 9, 9, 10; 1, 1), (8, 9, 9, 9.5; 0.9, 0.9)) |
| Hazard ID | Hazard-3 | Hazard-4 |
| Probability | ((3, 5, 5, 7; 1, 1), (4, 5, 5, 6; 0.9, 0.9)) | ((7, 9, 9, 10; 1, 1), (8, 9, 9, 9.5; 0.9, 0.9)) |
| Exposure | ((3, 5, 5, 7; 1, 1), (4, 5, 5, 6; 0.9, 0.9)) | ((7, 9, 9, 10; 1, 1), (8, 9, 9, 9.5; 0.9, 0.9)) |
| Consequence | ((5, 7, 7, 9; 1, 1), (6, 7, 7, 8; 0.9, 0.9)) | ((7, 9, 9, 10; 1, 1), (8, 9, 9, 9.5; 0.9, 0.9)) |
| Hazard ID | Hazard-5 | Hazard-6 |
| Probability | ((5, 7, 7, 9; 1, 1), (6, 7, 7, 8; 0.9, 0.9)) | ((3, 5, 5, 7; 1, 1), (4, 5, 5, 6; 0.9, 0.9)) |
| Exposure | ((7, 9, 9, 10; 1, 1), (8, 9, 9, 9.5; 0.9, 0.9)) | ((1, 3, 3, 5; 1, 1), (2, 3, 3, 4; 0.9, 0.9)) |
| Consequence | ((5, 7, 7, 9; 1, 1), (6, 7, 7, 8; 0.9, 0.9)) | ((5, 7, 7, 9; 1, 1), (6, 7, 7, 8; 0.9, 0.9)) |

In Step 5, the ranking value $Rank\left(\tilde{\tilde{R}}_{ij}\right)$ for each hazard is computed in $E_c$. In Step 6, there are $P_g = 720 (= 6!)$ permutations of the rankings must be tested for each of the hazards. In Step 7, the concordance and discordance index $I_j^l(H_\alpha, H_\beta)$ for each pair of hazard $(HR_\alpha, H_\beta)$ in the permutation $P_l$ for each criterion $c_i$ must be computed. For instance, the results of the concordance and discordance index $I_j^1(H1, H2)$ for seven combination of $P_{720}(H1, H2, H3, H4, H5, H6)$ permutation are shown in Table 8.5.

Then, we calculated $I_i^g(R_\alpha, R_\beta) \times W_j$ values and $I^g(R_\alpha, R_\beta)$ for each pair of $(R_\alpha, R_\beta)$ in $P_l$. In this step, we calculated $15 \times 720$ values of $I_i^g(R_\alpha, R_\beta).W_j$ for three parameters. In Step 9, the detailed concordance and discordance index was computed for each $P_g$. There are $P_g = 720 (= 6!)$ permutations. Furthermore, the ranking value $Rank(I^g)$ is calculated for the best 20 permutation $P_g$ that is presented in Table 8.6. Then, the ranking value is ordered in descending. The best permutation is $\mathrm{P}^{288} = (\mathrm{H}-4, \mathrm{H}-5, \mathrm{H}-1, \mathrm{H}-2, \mathrm{H}-3, \mathrm{H}-6)$. Results of the study demonstrate that the most important three risks are Flammable, flammable and explosive environments

**Table 8.5** The results of the concordance/discordance index for $P_{720}$

| | $I_j^{720}(H-1, H-2)$ | $I_j^{720}(H-1, H-3)$ | $I_j^{720}(H-1, H-4)$ | $I_j^{720}(H-1, H-5)$ | $I_j^{720}(H-1, H-6)$ |
|---|---|---|---|---|---|
| Probability | −11.50 | −22.46 | 0 | −11.50 | −11.50 |
| Exposure | −33.97 | −33.97 | −11.51 | 0 | −40.10 |
| Consequence | 0 | −10.96 | 0 | −10.96 | 11.50 |
| | $I_j^{720}(H-2, H-3)$ | $I_j^{720}(H-2, H-4)$ | $I_j^{720}(H-2, H-5)$ | $I_j^{720}(H-2, H-6)$ | $I_j^{720}(H-3, H-4)$ |
| Probability | −10.96 | 11.50 | 0 | 0 | 22.46 |
| Exposure | 0 | 22.46 | 33.97 | −6.14 | 22.46 |
| Consequence | −10.96 | 0 | −10.96 | 11.50 | 10.96 |
| | $I_j^{720}(H-3, H-5)$ | $I_j^{720}(H-3, H-6)$ | $I_j^{720}(H-4, H-5)-$ | $I_j^{720}(H-4, H-6)$ | $I_j^{720}(H-5, H-6)$ |
| Probability | 10.96 | 10.96 | −11.50 | −11.50 | 0 |
| Exposure | 33.97 | −6.14 | 11.51 | −28.60 | −40.11 |
| Consequence | 0 | 22.46 | −10.96 | 11.50 | 22.46 |

**Table 8.6** Final risk scores and rankings by the proposed approach

| | Permutations | Ranking |
|---|---|---|
| P288 | (H−4, H−5, H−1, H−2, H−3, H−6) | 867.317 |
| P282 | (H−4, H−5, H−2, H−1, H−3, H−6) | 837.184 |
| P348 | (H−4, H−1, H−5, H−2, H−3, H−6) | 830.406 |
| P287 | (H−4, H−5, H−1, H−2, H−6, H−3) | 825.125 |
| P286 | (H−4, H−5, H−1, H−3, H−2, H−6) | 809.913 |
| P281 | (H−4, H−5, H−2, H−1, H−6, H−3) | 794.992 |
| P347 | (H−4, H−1, H−5, H−2, H−6, H−3) | 788.214 |
| P346 | (H−4, H−1, H−5, H−3, H−2, H−6) | 773.001 |
| P324 | (H−4, H−2, H−5, H−1, H−3, H−6) | 770.139 |
| P168 | (H−5, H−4, H−1, H−2, H−3, H−6) | 769.249 |
| P358 | (H−4, H−1, H−2, H−5, H−3, H−6) | 763.361 |
| P280 | (H−4, H−5, H−2, H−3, H−1, H−6) | 749.645 |
| P162 | (H−5, H−4, H−2, H−1, H−3, H−6) | 739.116 |
| P334 | (H−4, H−2, H−1, H−5, H−3, H−6) | 733.227 |
| P323 | (H−4, H−2, H−5, H−1, H−6, H−3) | 727.947 |
| P167 | (H−5, H−4, H−1, H−2, H−6, H−3) | 727.057 |
| P284 | (H−4, H−5, H−1, H−6, H−2, H−3) | 725.529 |
| P276 | (H−4, H−5, H−3, H−1, H−2, H−6) | 722.374 |
| P357 | (H−4, H−1, H−2, H−5, H−6, H−3) | 721.169 |
| P166 | (H−5, H−4, H−1, H−3, H−2, H−6) | 711.844 |
| P285 | (H−4, H−5, H−1, H−3, H−6, H−2) | 710.316 |
| P161 | (H−5, H−4, H−2, H−1, H−6, H−3) | 696.924 |
| P660 | (H−1, H−4, H−5, H−2, H−3, H−6) | 695.426 |
| P274 | (H−4, H−5, H−3, H−2, H−1, H−6) | 692.241 |
| P333 | (H−4, H−2, H−1, H−5, H−6, H−3) | 691.035 |
| P344 | (H−4, H−1, H−5, H−6, H−2, H−3) | 688.617 |
| P322 | (H−4, H−2, H−5, H−3, H−1, H−6) | 682.600 |
| P345 | (H−4, H−1, H−5, H−3, H−6, H−2) | 673.404 |
| P283 | (H−4, H−5, H−1, H−6, H−3, H−2) | 668.124 |
| P278 | (H−4, H−5, H−2, H−6, H−1, H−3) | 665.261 |

(Hazard-4), Chemical Liquid + Gas (Alcohol, welding, soldering gas etc.) Exposure (Hazard-5), and Insufficient ventilation (Hazard-1).

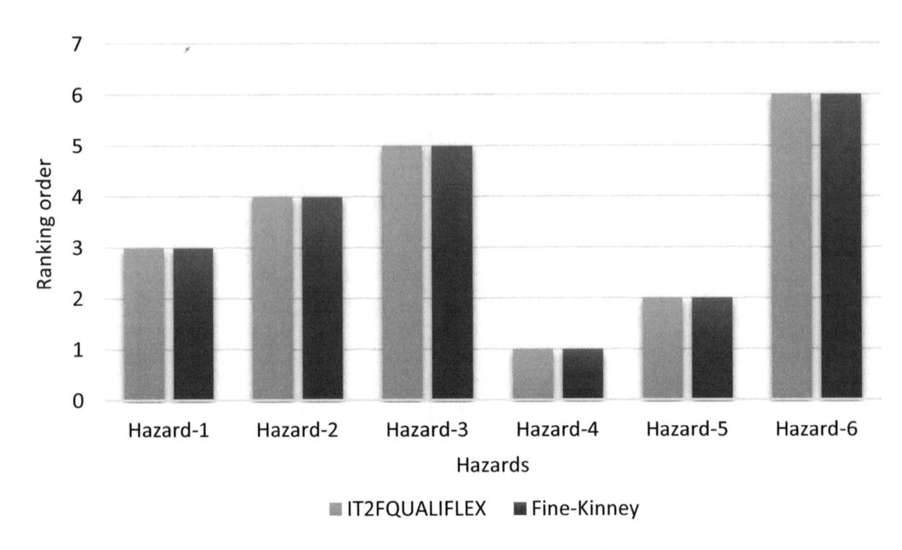


**Fig. 8.2** Comparison of rankings by proposed & classic approach

### *8.2.2 Validation Study on the Results*

In this subsection, some validation tests of the obtained ranking results are provided. As a first validation study, we made a comparative study between the results of the existed approach (IT2FQUALIFLEX under Fine–Kinney's method) and classical Fine–Kinney method. We then observe the variations in hazard rankings. The results are shown in Fig. 8.2.

It is observed from Fig. 8.2, that is obtained by both approaches, Hazard-4 is ranked as the most critical hazard, followed by Hazard-5. It is also seen that the least important three hazards (Hazard-2, Hazard-3, and Hazard-6) are the same according to both approaches.

## 8.3 Python Implementation of the Proposed Approach

```
# Chapter 8
# import required libraries
import numpy as np
import itertools as it
# set initial variables
n_criteria = 3
n_hazard = 6
nf_scores = 6 # number of linguistics
nf_values = 12 # number of fuzzy values
max_fuzzy_score = 10

criteria_weight = [0.289, 0.293, 0.418] # P, F, S
d = [1, 1, 0.9, 0.9] # degrees

fuzzy_scores = [
   ["P", 0, 1, 1, 3, d[0], d[1], 0.5, 1, 1, 2, d[2], d[3]], # Poor (P)
   ["MP", 1, 3, 3, 5, d[0], d[1], 2, 3, 3, 4, d[2], d[2]], # Medium Poor (MP)
   ["M", 3, 5, 5, 7, d[0], d[1], 4, 5, 5, 6, d[2], d[2]], # Medium (M)
   ["MG", 5, 7, 7, 9, d[0], d[1], 6, 7, 7, 8, d[2], d[2]], # Medium Good (MG)
   ["G", 7, 9, 9, 10, d[0], d[1], 8, 9, 9, 9.5, d[2], d[2]], # Good (G)
   ["VG", 9, 10, 10, 10, d[0], d[1], 9.5, 10, 10, 10, d[2], d[2]] # Very Good (VG)
]
# ehe: 1st expert's hazard evaluation
ehe = [
   # Hazard ID, Probability, Exposure,   Consequence
   ["Hazard-1", "MG", "VG", "M"],
   ["Hazard-2", "MG", "MP", "G"],
   ["Hazard-3", "M", "M", "MG"],
   ["Hazard-4", "G", "G", "G"],
   ["Hazard-5", "MG", "G", "MG"],
   ["Hazard-6", "M", "MP", "MG"]
]

def print_result(best_perm, perm_val):
   best_perm = np.asarray(best_perm)
   print('Hazard Id, Rank')
   for ix in range(0, len(best_perm)):
      print(ehe[ix][0], ', ', int(np.asarray(np.where(best_perm == ix))) + 1)
   print('The Best Permutation [Hazard-]', best_perm + 1)
   print('The Best Permutation Distance:', perm_val)
```

```
    # normalize fuzzy scores
    # nfs: normalized fuzzy scores
    nfs = [fs.copy() for fs in fuzzy_scores]
    for i in range(0, nf_scores):
        for j in range(1, nf_values + 1):
            if j == 5 or j == 6 or j == 11 or j == 12:
                continue
            nfs[i][j] /= max_fuzzy_score

    # initialize the matrix
    idm = [] # initial decision matrix
    for hz in ehe:
        temp_list_1 = []

    for cr in range(0, n_criteria):
        ix = 0
        for fs in fuzzy_scores:
            if hz[cr + 1] == fs[0]:
                temp_list_1.append(fs[1:])
            ix += 1
    idm.append(temp_list_1)

mean_val = np.zeros([n_hazard, n_criteria], dtype=float)
std_val = np.zeros([n_hazard, n_criteria], dtype=float)
membership_val = np.zeros([n_hazard, n_criteria], dtype=float)
for ix_hz in range(0, n_hazard):
    for ix_cr in range(0, n_criteria):
        for ix in [0, 1, 2, 6, 7, 8]:
            mean_val[ix_hz][ix_cr]+=np.mean(idm[ix_hz][ix_cr][ix+0:ix+2])
            std_val[ix_hz][ix_cr]+=np.std(idm[ix_hz][ix_cr][ix+0:ix+2])
        std_val[ix_hz][ix_cr] += np.std(idm[ix_hz][ix_cr][0:4])
        std_val[ix_hz][ix_cr] += np.std(idm[ix_hz][ix_cr][6:10])
        std_val[ix_hz][ix_cr] = std_val[ix_hz][ix_cr] / 4
        membership_val[ix_hz][ix_cr] = idm[ix_hz][ix_cr][4] \
                              + idm[ix_hz][ix_cr][5] \
                              + idm[ix_hz][ix_cr][10] \
                              + idm[ix_hz][ix_cr][11]

distance = mean_val - std_val + membership_val

comb = []
perm = list(it.permutations(range(0, n_hazard)))
for pr in perm:
    for cm in it.combinations(pr, 2):
        comb.append(list(cm))
```

```
perm_size = len(perm)
comb_size = len(comb)
# wpd: weighted pair distance
wpd = np.ones([n_criteria, comb_size, nf_values], dtype=float)
for cr in range(0, n_criteria):
   for cm in range(0, comb_size):
      delta_dist = distance[comb[cm][0]][cr] - distance[comb[cm][1]][cr]
      for ix in [0, 1, 2, 3, 6, 7, 8, 9]:
         wpd[cr][cm][ix] = delta_dist * criteria_weight[cr]
      counter = 0
      for ix in [4, 5, 10, 11]:
         wpd[cr][cm][ix] = d[counter]
         counter = counter + 1

# spd: sum pair distance
spd = np.sum(wpd, axis=0)
spd2 = np.reshape(spd, [perm_size, int(comb_size / perm_size), nf_values])
spd3 = np.sum(spd2, axis=1)
for pr in range(0, perm_size):
   counter = 0
   for ix in [4, 5, 10, 11]:
      spd3[pr][ix] = d[counter]
      counter = counter + 1

f_mean_val = np.zeros([perm_size, 1], dtype=float)
f_std_val = np.zeros([perm_size, 1], dtype=float)
f_membership_val = np.zeros([perm_size, 1], dtype=float)
for ix_pr in range(0, perm_size):
   for ix in [0, 1, 2, 6, 7, 8]:
      f_mean_val[ix_pr] += np.mean(spd3[ix_pr][ix + 0:ix + 2])
      f_std_val[ix_pr] += np.std(spd3[ix_pr][ix + 0:ix + 2])
   f_std_val[ix_pr] += np.std(spd3[ix_pr][0:4])
   f_std_val[ix_pr] += np.std(spd3[ix_pr][6:10])
   f_std_val[ix_pr] = f_std_val[ix_pr] / 4
   f_membership_val[ix_pr] = spd3[ix_hz][4] + spd3[ix_hz][5] \
                  + spd3[ix_hz][10] + spd3[ix_hz][11]
f_distance = f_mean_val - f_std_val + f_membership_val

best_ix = np.argmax(f_distance)
best = perm[best_ix]
print_result(best, f_distance[best_ix])

'''
Output:
Hazard Id, Rank
Hazard-1 , 3
Hazard-2 , 4
Hazard-3 , 5
Hazard-4 , 1
Hazard-5 , 2
Hazard-6 , 6
The Best Permutation [Hazard-] [4 5 1 2 3 6]
The Best Permutation Distance: [867.317432]
'''
```

# References


1. Paelinck, J. H. P. (1976). Qualitative multiple criteria analysis, environmental protection and multiregional development. *Papers of the Regional Science Association, 36,* 59–74.
2. Alinezhad, A., & Khalili, J. (2019). QUALIFLEX method. In *New methods and applications in multiple attribute decision making (MADM)* (pp. 41–46). Springer, Cham.
3. Chen, T. Y., Chang, C. H., & Lu, J. F. R. (2013). The extended QUALIFLEX method for multiple criteria decision analysis based on interval type-2 fuzzy sets and applications to medical decision making. *European Journal of Operational Research, 226*(3), 615–625.
4. Lee, L. W., Chen, S. M. (2008). Fuzzy multiple attributes group decision-making based on the extension of TOPSIS method and interval type-2 fuzzy sets. In *Proceedings of the seventh international conference on machine learning and cybernetic* (pp. 3260–3265). Taipei. Retrieved July 12–15, 2008.
5. Chen, S. M., & Lee, L. W. (2010). Fuzzy multiple attributes group decision-making based on the interval type-2 TOPSIS method. *Expert Systems with Applications, 37*(4), 2790–2798.
6. Gul, M., Guven, B., & Guneri, A. F. (2018). A new Fine–Kinney-based risk assessment framework using FAHP-FVIKOR incorporation. *Journal of Loss Prevention in the Process Industries, 53,* 3–16.
7. Celik, E., Aydin, N., & Gumus, A. T. (2014). A multiattribute customer satisfaction evaluation approach for rail transit network: A real case study for Istanbul, Turkey. *Transport Policy, 36,* 283–293.
8. Wang, J. C., Tsao, C. Y., & Chen, T. Y. (2015). A likelihood-based QUALIFLEX method with interval type-2 fuzzy sets for multiple criteria decision analysis. *Soft Computing, 19*(8), 2225–2243.
9. Zhang, X. (2016). Multicriteria Pythagorean fuzzy decision analysis: A hierarchical QUALIFLEX approach with the closeness index-based ranking methods. *Information Sciences, 330,* 104–124.
10. Chen, T. Y. (2014). Interval-valued intuitionistic fuzzy QUALIFLEX method with a likelihood-based comparison approach for multiple criteria decision analysis. *Information Sciences, 261,* 149–169.
11. Li, J., & Wang, J. Q. (2017). An extended QUALIFLEX method under probability hesitant fuzzy environment for selecting green suppliers. *International Journal of Fuzzy Systems, 19*(6), 1866–1879.
12. Zhang, X., & Xu, Z. (2015). Hesitant fuzzy QUALIFLEX approach with a signed distance-based comparison method for multiple criteria decision analysis. *Expert Systems with Applications, 42*(2), 873–884.
13. Zhang, X., Xu, Z., & Liu, M. (2016). Hesitant trapezoidal fuzzy QUALIFLEX method and its application in the evaluation of green supply chain initiatives. *Sustainability, 8*(9), 952.
14. Chen, T. Y. (2013). Data construction process and QUALIFLEX-based method for multiple-criteria group decision making with interval-valued intuitionistic fuzzy sets. *International Journal of Information Technology & Decision Making, 12*(03), 425–467.
15. Demirel, H., Akyuz, E., Celik, E., & Alarcin, F. (2019). An interval type-2 fuzzy QUALIFLEX approach to measure performance effectiveness of ballast water treatment (BWT) system on-board ship. *Ships and Offshore Structures, 14*(7), 675–683.
16. Gumus, A. T., Aydin, N., Celik, E. (2014). Passenger satisfaction evaluation for rail transit lines in Istanbul using qualiflex approach based on interval type-2 trapezoidal fuzzy numbers. In *CIE 2014—44th international conference on computers and industrial engineering and IMSS 2014—9th international symposium on intelligent manufacturing and service systems, joint international symposium on the social impacts of developments in information, manufacturing and service systems—proceedings* (pp. 343–357). Istanbul, Turkey.
17. Celik, E., Gul, M., Aydin, N., Gumus, A. T., & Guneri, A. F. (2015). A comprehensive review of multi criteria decision making approaches based on interval type-2 fuzzy sets. *Knowledge-Based Systems, 85,* 329–341.

18. Mendel, J. M., John, R. I., & Liu, F. L. (2006). Interval type-2 fuzzy logical systems made simple. *IEEE Transactions on Fuzzy Systems, 14*(6), 808–821.
19. Zhang, Z. (2017). Multi-criteria decision-making using interval-valued hesitant fuzzy QUALIFLEX methods based on a likelihood-based comparison approach. *Neural Computing and Applications, 28*(7), 1835–1854.

# Chapter 9
# Fine–Kinney-Based Occupational Risk Assessment Using Interval Type-2 Fuzzy VIKOR

**Abstract** In this chapter, we improved Fine–Kinney occupational risk assessment approach with interval type-2 fuzzy VIKOR (IT2FVIKOR). VIKOR is a compromise multi-attribute decision-making method proposed by Opricovic (1998). Similar to other compromised solution-based approaches, it considers the solution which is closest to the ideal. In this chapter, we adapted the interval type-2 fuzzy sets (IT2FSs) into VIKOR as it reflects the uncertainty well in decision-making. IT2FVIKOR algorithm under the Fine–Kinney concept provides a useful and solid approach to the occupational health and safety risk assessment. In addition to proposing this new approach, a case study is performed in a gun and rifle barrel external surface oxidation and coloring unit of a gun factory. A validation and a sensitivity analysis is also attached to this study. Finally, the proposed approach is implemented in Python.

## 9.1 Interval Type-2 Fuzzy VIKOR (IT2FVIKOR)

Prior to defining the proposed IT2FVIKOR, a general view on the classic VIKOR method is provided.

### *9.1.1 VIKOR Method*

The classical VIKOR method was originally suggested by [1]. It uses a parameter weight matrix, as well as a decision matrix, which consists of hazard, parameters, their respected performance measures.

The main steps of VIKOR consist of the following steps [2]:

1. Set up the problem and construct a decision matrix including alternatives and criteria,
2. Define the worst and the best values of all parameter functions,
3. Compute S and R values that are specific for VIKOR,
4. Compute Q values according to the computed S and R values from Step 2,



M. Gul et al., *Fine–Kinney-Based Fuzzy Multi-criteria Occupational Risk Assessment*, Studies in Fuzziness and Soft Computing 398,
https://doi.org/10.1007/978-3-030-52148-6_9

5. Rank the hazards using the value of S, R, and Q in decreasing order
6. Suggest a compromise solution provided two situations (acceptable advantage and acceptable stability) are satisfied.

### 9.1.2 Proposed Fine–Kinney-Based Approach Using IT2FVIKOR

IT2FVIKOR is based on a compromise solution [1, 3, 4] that provides a maximum group utility two-tuple fuzzy numbers [8], Interval-valued fuzzy sets [5], triangular intuitionistic fuzzy numbers [7], and interval–valued fuzzy with grey relational analysis [6], are integrated with VIKOR. Since type-2 fuzzy sets have additional degrees of freedom, they reflect more uncertainty than type-1 fuzzy sets. [9–12]. In this chapter, the IT2FVIKOR is proposed to prioritize hazards and associated risks in the OHS perspective. In the OHS risk assessment process, it is assumed that there are $m$ hazards (emerged hazards and associated risks), where $\{R_1, R_2, \ldots, R_m\}$, $n$ parameters (risk parameters of Fine–Kinney concept) $\{A_1, A_2, \ldots, A_n\}$ and $L$ OHS experts E $\{E_1, E_2, \ldots, E_L\}$.

The steps of the IT2FSVIKOR are given in detail as follows [13]:

**Step 1**: The importance weights of the parameters are computed using Eq. (9.1).

$$A_n = \left(\tilde{\tilde{a}}_j^k\right)_{nx1} = \begin{matrix} A_1 \\ A_2 \\ \vdots \\ A_n \end{matrix} \begin{bmatrix} \tilde{\tilde{a}}_1^k \\ \tilde{\tilde{a}}_2^k \\ \ldots \\ \tilde{\tilde{a}}_n^k \end{bmatrix} \tag{9.1}$$

where $\tilde{\tilde{a}}_j = \left(\frac{\tilde{\tilde{a}}_j^1 \oplus \tilde{\tilde{a}}_j^2 \oplus \cdots \oplus \tilde{\tilde{a}}_j^L}{L}\right)$, $\tilde{\tilde{a}}_j$ is an IT2FS $1 \le i \le m,\ 1 \le j \le n,\ 1 \le k \le L$.

**Step 2**: The average IT2F values of hazards are also calculated using Eq. (9.2).

$$E_c = (\tilde{\tilde{e}}_{ij}^k)_{n \times m} = \begin{matrix} & R_1 & R_2 & \cdots & R_m \\ A_1 & \tilde{\tilde{e}}_{11}^k & \tilde{\tilde{e}}_{12}^k & \cdots & \tilde{\tilde{e}}_{1m}^k \\ A_2 & \tilde{\tilde{e}}_{21}^k & \tilde{\tilde{e}}_{22}^k & \cdots & \tilde{\tilde{e}}_{2m}^k \\ \vdots & \vdots & \vdots & \vdots & \vdots \\ A_n & \tilde{\tilde{e}}_{n1}^k & \tilde{\tilde{e}}_{n2}^k & \cdots & \tilde{\tilde{e}}_{nm}^k \end{matrix} \tag{9.2}$$

where $\tilde{\tilde{e}}_{ij} = \left(\frac{\tilde{\tilde{e}}_{ij}^1 \oplus \tilde{\tilde{e}}_{ij}^2 \oplus \cdots \oplus \tilde{\tilde{e}}_{ij}^L}{L}\right)$.

**Step 3**: The weighted IT2F decision matrix is calculated as follows:

$$\tilde{\tilde{V}} = \left[\tilde{\tilde{v}}_{ij}\right]_{mxn} \tag{9.3}$$

where

$$\tilde{\tilde{v}}_{ij} = \tilde{\tilde{a}}_j \otimes \tilde{\tilde{e}}_{ij} = \begin{pmatrix} \left(f_{i1}^U, f_{i2}^U, f_{i3}^U, f_{i4}^U; H_1\left(\tilde{F}_i^U\right), H_2\left(\tilde{F}_i^U\right)\right), \\ \left(f_{i1}^L, f_{i2}^L, f_{i3}^L, f_{i4}^L; H_1\left(\tilde{F}_i^L\right), H_2\left(\tilde{F}_i^L\right)\right) \end{pmatrix}$$

**Step 4**: The negative ($N^{e-}$) and positive ($P^{e*}$,$P^{v*}$) ideal solution for lower and upper reference points of the IT2FNs are calculated [14].

$$N^{e-} = \left\{\tilde{\tilde{e}}_{ij}, \tilde{\tilde{e}}_{ij}, \ldots, \tilde{\tilde{e}}_{ij}\right\} = \left\{\min_i \tilde{\tilde{e}}_{ij} \middle| j \in Benefit\right\}$$

$$N^{e-} = \begin{pmatrix} \left(e_{i1}^{U-}, e_{i2}^{U-}, e_{i3}^{U-}, e_{i4}^{U-}; \min H_1\left(\tilde{E}_i^U\right), \min H_2\left(\tilde{E}_i^U\right)\right), \\ \left(e_{i1}^{L-}, e_{i2}^{L-}, e_{i3}^{L-}, e_{i4}^{L-}; \min H_1\left(\tilde{E}_i^L\right), \min H_2\left(\tilde{E}_i^L\right)\right) \end{pmatrix}$$

$$P^{e*} = \left\{\tilde{\tilde{e}}_{ij}^*, \tilde{\tilde{e}}_{ij}^*, \ldots, \tilde{\tilde{e}}_{ij}^*\right\} = \left\{\max_i \tilde{\tilde{e}}_{ij} \middle| j \in Benefit\right\}$$

$$P^{e*} = \begin{pmatrix} \left(e_{i1}^{U*}, e_{i2}^{U*}, e_{i3}^{U*}, e_{i4}^{U*}; \max H_1\left(\tilde{E}_i^U\right), \max H_2\left(\tilde{E}_i^U\right)\right), \\ \left(e_{i1}^{L*}, e_{i2}^{L*}, e_{i3}^{L*}, e_{i4}^{L*}; \max H_1\left(\tilde{E}_i^L\right), \max H_2\left(\tilde{E}_i^L\right)\right) \end{pmatrix}$$

$$P^{v*} = \left\{\tilde{\tilde{v}}_{ij}^*, \tilde{\tilde{v}}_{ij}^*, \ldots, \tilde{\tilde{v}}_{ij}^*\right\} = \left\{\max_i \tilde{\tilde{v}}_{ij} \middle| j \in Benefit\right\}$$

$$P^{v*} = \begin{pmatrix} \left(f_{i1}^{U*}, f_{i2}^{U*}, f_{i3}^{U*}, f_{i4}^{U*}; \max H_1\left(\tilde{F}_i^U\right), \max H_2\left(\tilde{F}_i^U\right)\right), \\ \left(f_{i1}^{L*}, f_{i2}^{L*}, f_{i3}^{L*}, f_{i4}^{L*}; \max H_1\left(\tilde{F}_i^L\right), \max H_2\left(\tilde{F}_i^L\right)\right) \end{pmatrix}$$

Next, the worst ($R_i$) and the average ($S_i$)group scores for each hazard is calculated.

$$R_i = \max_j\left(\frac{1}{2}\left(S_{ij}^U + S_{ij}^L\right)\right), \ \forall i = 1, \ldots, m \tag{9.4}$$

$$S_i = \sum_{j=1}^{n} \frac{1}{2}\left(S_{ij}^U + S_{ij}^L\right), \ \forall i = 1, \ldots, m \tag{9.5}$$

where

$$S_{ij}^{U} = \sum_{j} \frac{\sqrt{\frac{1}{4}\sum_{k=1}^{4}\left[\left(f_{i1}^{U*} - f_{i4}^{U}\right)^{2} + \left(f_{i2}^{U*} - f_{i3}^{U}\right)^{2} + \left(f_{i3}^{U*} - f_{i2}^{U}\right)^{2} + \left(f_{i4}^{U*} - f_{i1}^{U}\right)^{2}\right]}}{\sqrt{\frac{1}{4}\sum_{k=1}^{4}\left[\left(e_{i1}^{U*} - e_{i4}^{U-}\right)^{2} + \left(e_{i2}^{U*} - e_{i3}^{U-}\right)^{2} + \left(e_{i3}^{U*} - e_{i2}^{U-}\right)^{2} + \left(e_{i4}^{U*} - e_{i1}^{U-}\right)^{2}\right]}},$$

$\forall i = 1, \ldots, m$

$$S_{ij}^{L} = \sum_{j} \frac{\sqrt{\frac{1}{4}\sum_{k=1}^{4}\left[\left(f_{i1}^{L*} - f_{i4}^{L}\right)^{2} + \left(f_{i2}^{L*} - f_{i3}^{L}\right)^{2} + \left(f_{i3}^{L*} - f_{i2}^{L}\right)^{2} + \left(f_{i4}^{L*} - f_{i1}^{U}\right)^{2}\right]}}{\sqrt{\frac{1}{4}\sum_{k=1}^{4}\left[\left(e_{i1}^{L*} - e_{i4}^{L-}\right)^{2} + \left(e_{i2}^{L*} - e_{i3}^{L-}\right)^{2} + \left(e_{i3}^{L*} - e_{i2}^{L-}\right)^{2} + \left(e_{i4}^{L*} - e_{i1}^{L-}\right)^{2}\right]}},$$

$\forall i = 1, \ldots, m$

**Step 5**: The $Q_i$ is calculated according to the $S_i$ and $R_i$ using Eq. (9.6).

$$Q_i = v\frac{(S_i - S^*)}{(S^- - S^*)} + (1 - v)\frac{(R_i - R^*)}{(R^* - R^*)} \tag{9.6}$$

where $S^* = \min_i S_i$, $S^- = \max_i S_i$, $R^* = \min_i R_i$, $R^- = \max_i R_i$, and $v \in [0, 1]$ is the weight of the decision-making strategy of the maximum group utility. The smallest $Q_i$ is then selected as a compromise solution if two conditions are acceptable [1].

Figure 9.1 presents the main steps of the proposed approach.

*Step 1*
Computation of the importance weights of the Fine Kinney risk parameters

*Step 2*
Computation of the average fuzzy performance values of hazards

*Step 3*
Computation of the weighted type-2 fuzzy decision matrix

*Step 4*
Computation of the average (S) and the worst (R) group scores for each hazard

*Step 5*
Computation of the Q values for each hazard according to the S and R values computed in Step 4.

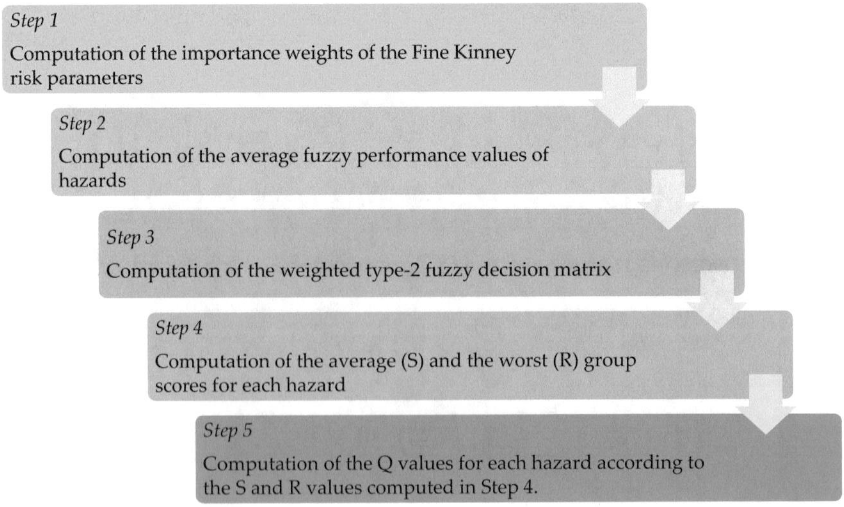

**Fig. 9.1** The main steps of the proposed approach

## 9.2 Case Study

In applying the proposed approach, we demonstrate a case study for the occupational risk assessment of a gun and rifle barrel external surface oxidation and coloring unit of a gun factory. The case study is previously studied in [15]. However, classical type-1 fuzzy sets were integrated to VIKOR in that study. Also, the current study is differentiated from [15], by the aspects of covering different risk parameters. In [15], Fine–Kinney concept is not taken into consideration. Therefore, the existed work is indeed novel in terms of methodology and not the same of [15], in terms of case study demonstration scope and boundaries. As the main method of the approach, IT2FVIKOR was implemented for ranking the hazards and their related risks. In the following, the stepwise application of the proposed approach to the problem is provided.

### *9.2.1 Application Results*

In this application, three parameters of Fine–Kinney as probability ($P$), exposure ($E$), and consequence ($C$) are used in assessing hazards. Also, the risk list regarding the gun and rifle barrel external surface oxidation and coloring unit of the gun factory is provided in Table 9.1. The weights of these parameters are derived from [16], as $w_P = 0.289$, $w_E = 0.293$, $w_C = 0.418$. During the prioritization procedure by IT2FVIKOR, the used linguistic terms are given in Table 9.2. This scale is obtained from Celik et al. [17]. While using the terms from the table, the evaluations given in Table 9.3, are used to obtain the fuzzy aggregated decision matrix.

By following the steps as indicated in Sect. 9.1, we obtain the final values of IT2FVIKOR model. Using the model, $S_i$, $R_i$, and $Q_i$ values are calculated and hazards are ranked according to the increasing order of $Q_i$ values (Table 9.4). Results of the study demonstrate that the most critical three risks are flammable and explosive environments (Hazard-16), insufficient ventilation (Hazard-4), and noise (Hazard-13).

### *9.2.2 Validation Study on the Results*

In this subsection, some validation tests of the obtained ranking results are provided. As a first validation study, we made a comparative study between the results of the existed approach (IT2FVIKOR under Fine–Kinney's method) and classical Fine–Kinney method. We then observe the variations in hazard rankings. The results are shown in Fig. 9.2.

It is observed from Fig. 9.2 that by both approaches, Hazard-16 is ranked as the most critical hazard, followed by Hazard-4. It is also seen that the least important

**Table 9.1** The hazard identification list in chrome plating unit

| ID | Hazard identification |
|---|---|
| Hazard-1 | Layout of the working area (tripping, falling) |
| Hazard-2 | Wet floor (falling, slipping, tripping) |
| Hazard-3 | Lifting, manual handling, pushing, placing, loading, |
| Hazard-4 | Insufficient ventilation |
| Hazard-5 | Continuous standing and settlement |
| Hazard-6 | Working with hammer, screwdriver, scissors, grinder |
| Hazard-7 | Repetitive movements |
| Hazard-8 | Rotating-moving parts of machines and its components (hit, squeeze, crush) |
| Hazard-9 | Electricity |
| Hazard-10 | Falling/flying objects |
| Hazard-11 | Fire |
| Hazard-12 | Inappropriate climatic environments (heat, wind, ice, gusts, hail, storm, cold) |
| Hazard-13 | Noise |
| Hazard-14 | Working with crush, squeeze and hit |
| Hazard-15 | Brightness, more lighting, inadequate lighting |
| Hazard-16 | Flammable and explosive environments |
| Hazard-17 | Exposure to chemical fluid, alcohol, welding and soldering gas |
| Hazard-18 | Emergency situations (fire, earthquake, flood etc.) |
| Hazard-19 | Water tank, pool |
| Hazard-20 | Professional competence/experience |

**Table 9.2** The scale used in assessing hazards

| Linguistic term | Interval type-2 fuzzy numbers |
|---|---|
| Poor (P) | ((0, 1, 1, 3; 1, 1), (0.5, 1, 1, 2; 0.9, 0.9)) |
| Medium Poor (MP) | ((1, 3, 3, 5; 1, 1), (2, 3, 3, 4; 0.9, 0.9)) |
| Medium (M) | ((3, 5, 5, 7; 1, 1), (4, 5, 5, 6; 0.9, 0.9)) |
| Medium Good (MG) | ((5, 7, 7, 9; 1, 1), (6, 7, 7, 8; 0.9, 0.9)) |
| Good (G) | ((7, 9, 9, 10; 1, 1), (8, 9, 9, 9.5; 0.9, 0.9)) |
| Very Good (VG) | ((9, 10, 10, 10; 1, 1), (9.5, 10, 10, 10; 0.9, 0.9)) |

three hazards (Hazard-12, Hazard-10, and Hazard-15) are the same according to both approaches. When we compare the results obtained by both approaches, we observe that there are very small rank variations between them. The Spearman rank correlation between the two approaches is obtained as 0.948. That means there exists a high correlation between the ranking orders of two approaches. So that, it can be

**Table 9.3** The linguistic initial decision matrix from the OHS experts' consensus

| Hazard ID | Probability | Exposure | Consequence |
|---|---|---|---|
| Hazard-1 | M | VG | M |
| Hazard-2 | G | VG | M |
| Hazard-3 | MG | VG | M |
| Hazard-4 | G | VG | MG |
| Hazard-5 | MG | VG | M |
| Hazard-6 | MG | G | M |
| Hazard-7 | MG | VG | M |
| Hazard-8 | M | MP | G |
| Hazard-9 | M | P | G |
| Hazard-10 | M | P | M |
| Hazard-11 | M | VG | MG |
| Hazard-12 | M | MP | MP |
| Hazard-13 | MG | VG | MG |
| Hazard-14 | MG | G | M |
| Hazard-15 | M | MP | M |
| Hazard-16 | MG | VG | VG |
| Hazard-17 | MG | VG | MG |
| Hazard-18 | M | MG | MG |
| Hazard-19 | MG | VG | M |
| Hazard-20 | MG | MG | MG |

**Table 9.4** Final risk scores and rankings by the proposed approach

| Hazard | $S_i$ | $R_i$ | $Q_i$(v = 0.5) | Rank | Hazard | $S_i$ | $R_i$ | $Q_i$(v = 0.5) | Rank |
|---|---|---|---|---|---|---|---|---|---|
| Hazard-1 | 0.608 | 0.301 | 0.524 | 11 | Hazard-11 | 0.496 | 0.289 | 0.420 | 6 |
| Hazard-2 | 0.425 | 0.301 | 0.396 | 5 | Hazard-12 | 0.941 | 0.418 | 1.000 | 15 |
| Hazard-3 | 0.497 | 0.301 | 0.446 | 7 | Hazard-13 | 0.384 | 0.189 | 0.134 | 3 |
| Hazard-4 | 0.313 | 0.189 | 0.083 | 2 | Hazard-14 | 0.529 | 0.301 | 0.468 | 8 |
| Hazard-5 | 0.497 | 0.301 | 0.446 | 7 | Hazard-15 | 0.824 | 0.301 | 0.676 | 13 |
| Hazard-6 | 0.529 | 0.301 | 0.468 | 8 | Hazard-16 | 0.227 | 0.178 | 0.000 | 1 |
| Hazard-7 | 0.497 | 0.301 | 0.446 | 7 | Hazard-17 | 0.384 | 0.189 | 0.134 | 3 |
| Hazard-8 | 0.612 | 0.289 | 0.501 | 10 | Hazard-18 | 0.584 | 0.289 | 0.481 | 9 |
| Hazard-9 | 0.671 | 0.293 | 0.551 | 12 | Hazard-19 | 0.497 | 0.301 | 0.446 | 7 |
| Hazard-10 | 0.883 | 0.301 | 0.717 | 14 | Hazard-20 | 0.472 | 0.189 | 0.195 | 4 |

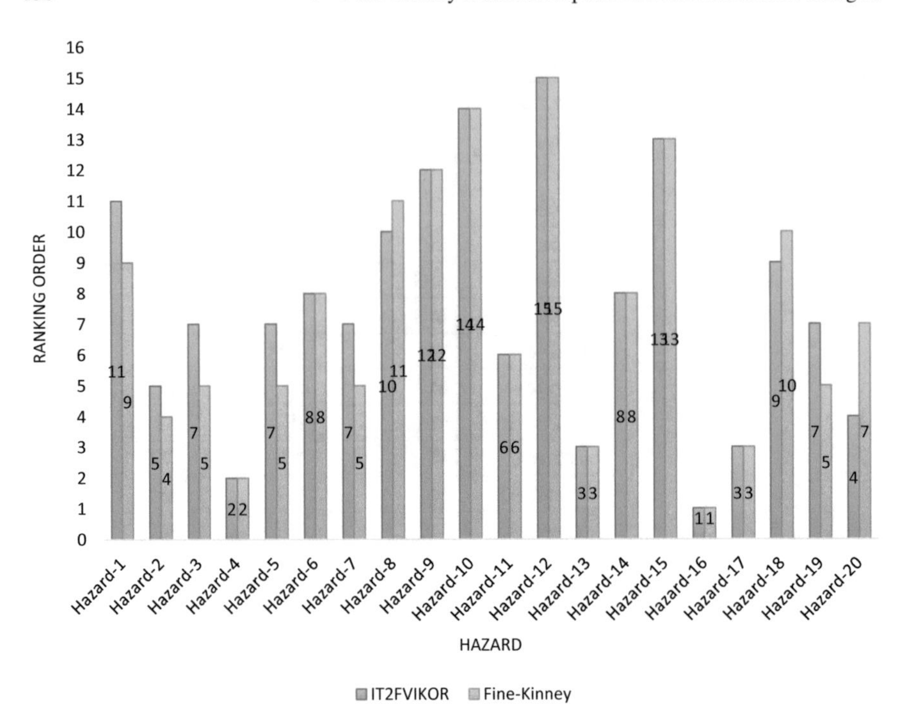


**Fig. 9.2** Comparison of rankings by proposed & classic approach

claimed that this proposed approach is applicable for occupational risk assessment in the Fine–Kinney domain.

A sensitivity analysis was also implemented in the results of the proposed approach by the varying value of $v$ as a secondary validation study. The $v$ value was taken as 0.5 in the above provided results. The $v$ value was analyzed starting from 0 to 1 with an increment of 0.1. Therefore, eleven different scenarios are investigated to observe the variability. This analysis aims to explain how rankings of hazards are changing according to these different scenarios. Q value results from the sensitivity analysis are given in Fig. 9.3.

According to the observation of Fig. 9.3; the Hazard-16 is determined as the most critical hazard for all case. We also analyzed the first most critical three hazards, the following results are obtained:

- When the $v$ value is greater than 0.5, the trend where either in increasing or decreasing is erratic for Hazard-3. Whereas the trend is not erratic for all hazards if the $v$ values are equal to 0.4 or less than 0.4.
- It is observed that the priority of the most critical hazard (Hazard-16) and least serious hazard (Hazard-12) are constant with respect to $Q$ values with different $v$ values.

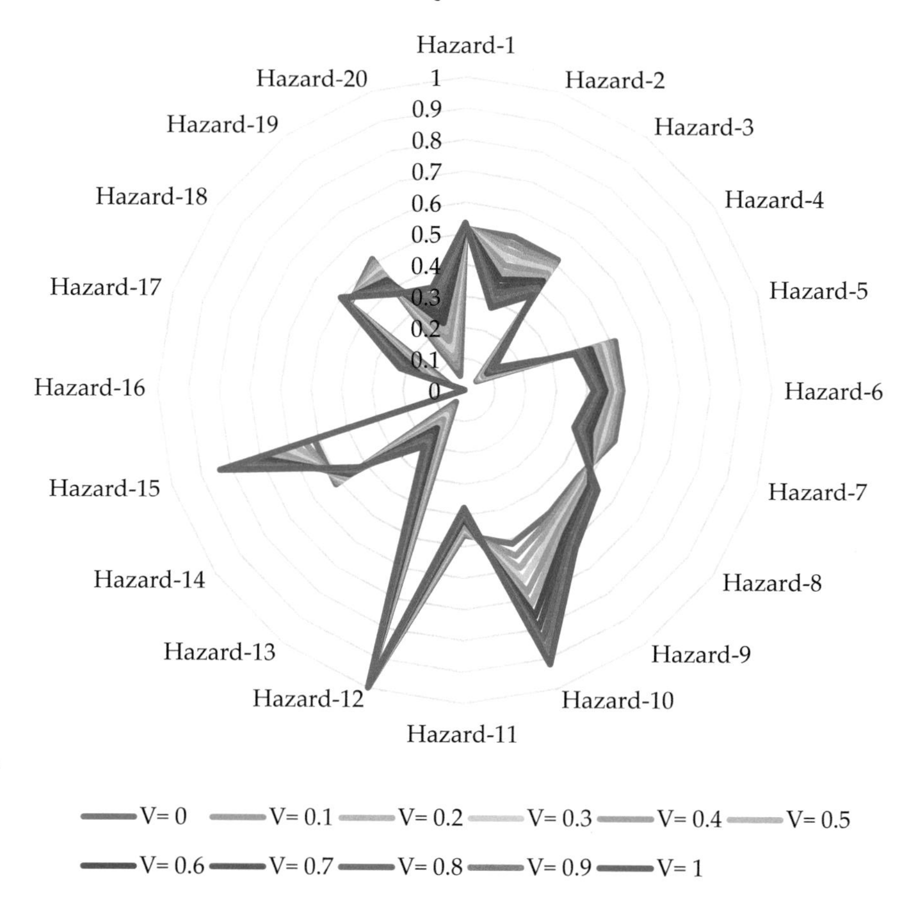


**Fig. 9.3** IT2FVIKOR Q values according to different $v$ values

- If decision-makers are in consensus that means $v$ is equal to 0.5, the final ranking of hazards is determined as $Hazard-16 \succ Hazard-4 \succ Hazard-13 \succ \cdots \succ Hazard-15 \succ Hazard-10 \succ Hazard-12$.

  This chapter confirms that the results in the ranks of risk are consistent.

## 9.3 Python Implementation of the Proposed Approach

```
# Chapter 9
# import required libraries
import numpy as np

# set initial variables
n_criteria = 3
n_hazards = 20
nf_scores = 6 # number of linguistics
nf_values = 12 # number of fuzzy values
max_fuzzy_score = 10

criteria_weight = [0.289, 0.293, 0.418] # P, F, S

fuzzy_scores = [["P", 0, 1, 1, 3, 1, 1, 0.5, 1, 1, 2, 0.9, 0.9], # Poor (P)
            ["MP", 1, 3, 3, 5, 1, 1, 2, 3, 3, 4, 0.9, 0.9], # Medium Poor (MP)
            ["M", 3, 5, 5, 7, 1, 1, 4, 5, 5, 6, 0.9, 0.9], # Medium (M)
            ["MG", 5, 7, 7, 9, 1, 1, 6, 7, 7, 8, 0.9, 0.9], # Medium Good (MG)
            ["G", 7, 9, 9, 10, 1, 1, 8, 9, 9, 9.5, 0.9, 0.9], # Good (G)
            ["VG", 9, 10, 10, 10, 1, 1, 9.5, 10, 10, 10, 0.9, 0.9] # Very Good (VG)
            ]

# ehe: 1st expert's hazard evaluation
ehe = [ # Hazard ID, Probability, Exposure,  Consequence
  ["Hazard-1", "M", "VG", "M"],
  ["Hazard-2", "G", "VG", "M"],
  ["Hazard-3", "MG", "VG", "M"],
  ["Hazard-4", "G", "VG", "MG"],
  ["Hazard-5", "MG", "VG", "M"],
  ["Hazard-6", "MG", "G", "M"],
  ["Hazard-7", "MG", "VG", "M"],
  ["Hazard-8", "M", "MP", "G"],
  ["Hazard-9", "M", "P", "G"],
  ["Hazard-10", "M", "P", "M"],
  ["Hazard-11", "M", "VG", "MG"],
  ["Hazard-12", "M", "MP", "MP"],
  ["Hazard-13", "MG", "VG", "MG"],
  ["Hazard-14", "MG", "G", "M"],
  ["Hazard-15", "M", "MP", "M"],
  ["Hazard-16", "MG", "VG", "VG"],
  ["Hazard-17", "MG", "VG", "MG"],
  ["Hazard-18", "M", "MG", "MG"],
  ["Hazard-19", "MG", "VG", "M"],
  ["Hazard-20", "MG", "MG", "MG"]
]
```

```
def rank(vector, da): # da -1:descending, 1:ascending
    order = np.zeros([len(vector), 1])
    unique_val = da * np.sort(da * np.unique(vector))
    for ix in range(0, len(unique_val)):
        order[np.argwhere(vector == unique_val[ix])] = ix + 1
    return order


def print_result(order, vector):
    print('Hazard Id, Rank, Value')
    for ix in range(0, len(order)):
        print(ehe[ix][0], ', ', int(order[ix]), ', ', vector[ix])


# initilize the matrixes
idm = [] # initial_decision_matrix
ndm = [] # normalized_decision_matrix

# normalize fuzzy scores
# nfs: normalized fuzzy scores
nfs = [fs.copy() for fs in fuzzy_scores]
for i in range(0, nf_scores):
    for j in range(1, nf_values + 1):
        if j == 5 or j == 6 or j == 11 or j == 12:
            continue
        nfs[i][j] /= max_fuzzy_score

# set initial matrices
for hz in ehe:
    temp_list_1 = []
    temp_list_2 = []
    for cr in range(0, n_criteria):
        ix = 0
        for fs in fuzzy_scores:
            if hz[cr + 1] == fs[0]:
                temp_list_1.append(fs[1:])
                temp_list_2.append(nfs[ix][1:])
            ix += 1
    idm.append(temp_list_1)
    ndm.append(temp_list_2)

idm_max, idm_min = np.max(idm, axis=0), np.min(idm, axis=0)
idm_max = np.reshape(idm_max, [n_criteria, 2, int(nf_values / 2)])[:, :, 0:4]
idm_min = np.reshape(idm_min, [n_criteria, 2, int(nf_values / 2)])[:, :, 0:4]
reversed_idm_min = np.flip(idm_min, axis=2)
max_min_val = np.sqrt(np.sum(np.square(idm_max - reversed_idm_min), axis=2) / 4)
```

```
# wdm: weighted decision matrix
wdm = np.zeros([n_hazards, n_criteria, nf_values])
for ix_h in range(0, n_hazards):
    for ix_c in range(0, n_criteria):
        for ix_f in range(0, nf_values):
            if ix_f == 4 or ix_f == 5 or ix_f == 10 or ix_f == 11:
                continue
            wdm[ix_h][ix_c][ix_f] = idm[ix_h][ix_c][ix_f] * criteria_weight[ix_c]
w_max, w_min = np.max(wdm, axis=0), np.min(wdm, axis=0)
w_max = np.reshape(w_max, [n_criteria, 2, int(nf_values / 2)])[:, :, 0:4]
reversed_w_max = np.flip(w_max, axis=2)
r_wdm = np.reshape(wdm, [n_hazards, n_criteria, 2, int(nf_values / 2)])[:, :, :, 0:4]
si = []
for vl in r_wdm:
    temp_val = np.sqrt(
        np.sum(np.square(vl - reversed_w_max), axis=2) / 4)
    si.append(temp_val / max_min_val)
si = np.average(si, axis=2)
sum_si, max_si = np.sum(si, axis=1), np.max(si, axis=1)
mx_sum_si, mn_sum_si = np.max(sum_si), np.min(sum_si)
mx_max_si, mn_max_si = np.max(max_si), np.min(max_si)
weight = 0.5
q_val = weight * (sum_si - mn_sum_si) / (mx_sum_si - mn_sum_si) \
    + (1 - weight) * (max_si - mn_max_si) / (mx_max_si - mn_max_si)
hazard_rank = rank(q_val, 1)
print_result(hazard_rank, q_val)
'''
Output:
Hazard Id, Rank, Value
Hazard-1 , 11 , 0.5241441961634641
Hazard-2 , 5 , 0.3958161918294625
Hazard-3 , 7 , 0.4460858697902135
Hazard-4 , 2 , 0.083339275918433
Hazard-5 , 7 , 0.4460858697902135
Hazard-6 , 8 , 0.4684490746891673
Hazard-7 , 7 , 0.4460858697902135
Hazard-8 , 10 , 0.5011592880362385
Hazard-9 , 12 , 0.5506904086795759
Hazard-10 , 14 , 0.7168235058089443
Hazard-11 , 6 , 0.4198235445368992
Hazard-12 , 15 , 1.0
Hazard-13 , 3 , 0.13360895387918395
Hazard-14 , 8 , 0.4684490746891673
Hazard-15 , 13 , 0.675610164656116
Hazard-16 , 1 , 0.0
Hazard-17 , 3 , 0.13360895387918395
Hazard-18 , 9 , 0.48143652664555686
Hazard-19 , 7 , 0.4460858697902135
Hazard-20 , 4 , 0.19522193598784165
'''
```

## References

1. Opricovic, S. (1998). *Multicriteria optimization of civil engineering systems*. Belgrade: Faculty of Civil Engineering.
2. Gul, M., Celik, E., Aydin, N., Gumus, A. T., & Guneri, A. F. (2016). A state of the art literature review of VIKOR and its fuzzy extensions on applications. *Applied Soft Computing, 46,* 60–89.
3. Opricovic, S., & Tzeng, G. H. (2004). Compromise solution by MCDM methods: A comparative analysis of VIKOR and TOPSIS. *European Journal of Operational Research, 156*(2), 445–455.
4. Tzeng, G. H., Lin, C. W., & Opricovic, S. (2005). Multi-criteria analysis of alternative-fuel buses for public transportation. *Energy Policy, 33*(11), 1373–1383.
5. Ju, Y., & Wang, A. (2012). Extension of VIKOR method for multi-criteria group decision making problem with linguistic information. *Applied Mathematical Modelling, 37,* 3112–3125.
6. Vahdani, B., Hadipour, H., Sadaghiani, J. S., & Amiri, M. (2010). Extension of VIKOR method based on interval-valued fuzzy sets. *The International Journal of Advanced Manufacturing Technology, 47*(9–12), 1231–1239.
7. Wan, S. P., Wang, Q. Y., & Dong, J. Y. (2013). The extended VIKOR method for multi-attribute group decision making with triangular intuitionistic fuzzy numbers. *Knowledge-Based Systems, 52,* 65–77.
8. Kuo, M. S. (2011). A novel interval-valued fuzzy MCDM method for improving airlines' service quality in Chinese cross-strait airlines. *Transportation Research Part E: Logistics and Transportation Review, 47*(6), 1177–1193.
9. Chen, S. M., & Lee, L. W. (2010). Fuzzy multiple attributes group decision-making based on the interval type-2 TOPSIS method. *Expert Systems with Applications, 37*(4), 2790–2798.
10. Celik, E., & Gumus, A. T. (2016). An outranking approach based on interval type-2 fuzzy sets to evaluate preparedness and response ability of non-governmental humanitarian relief organizations. *Computers & Industrial Engineering, 101,* 21–34.
11. Celik, E. (2017). A cause and effect relationship model for location of temporary shelters in disaster operations management. *International Journal of Disaster Risk Reduction, 22,* 257–268.
12. Celik, E., & Gumus, A. T. (2018). An assessment approach for non-governmental organizations in humanitarian relief logistics and an application in Turkey. *Technological and Economic Development of Economy, 24*(1), 1–26.
13. Celik, E., Gul, M., Aydin, N., Gumus, A. T., & Guneri, A. F. (2015). A comprehensive review of multi criteria decision making approaches based on interval type-2 fuzzy sets. *Knowledge-Based Systems, 85,* 329–341.
14. Kuo, M. S., & Liang, G. S. (2012). A soft computing method of performance evaluation with MCDM based on interval-valued fuzzy numbers. *Applied Soft Computing, 12*(1), 476–485.
15. Gul, M. (2018). Application of Pythagorean fuzzy AHP and VIKOR methods in occupational health and safety risk assessment: The case of a gun and rifle barrel external surface oxidation and colouring unit. *International journal of occupational safety and ergonomics*, 1–14.
16. Gul, M., Guven, B., & Guneri, A. F. (2018). A new Fine–Kinney-based risk assessment framework using FAHP-FVIKOR incorporation. *Journal of Loss Prevention in the Process Industries, 53,* 3–16.
17. Celik, E., Aydin, N., & Gumus, A. T. (2014). A multiattribute customer satisfaction evaluation approach for rail transit network: A real case study for Istanbul, Turkey. *Transport Policy, 36,* 283–293.

Printed in the United States
by Baker & Taylor Publisher Services